EUGENICS AND POLITICS IN BRITAIN, 1900–1914

SCIENCE IN HISTORY

3

EUGENICS AND POLITICS IN BRITAIN, 1900–1914

EUGENICS AND POLITICS IN BRITAIN 1900–1914

by

G. R. SEARLE

Leyden
Noordhoff International Publishing
1976

A division of A. W. Sijthoff International Publishing Company B.V.
Leyden, The Netherlands

ISBN 90 286 0236 4

Set-in-type in the United Kingdom
Printed in The Netherlands

Contents

Acknowledgement

I am grateful to the Council of the Eugenics Society for the help they have given me, and for permitting me to see and to quote from the Society's Minute Books.

Introduction

Eugenics, to quote the definition of the man who coined the word, Francis Galton, is 'the study of agencies under social control that may improve or impair the racial qualities of future generations either physically or mentally'.[1] Eugenists believe that the knowledge so acquired can be applied to the practical task of raising the level of fitness in the human race. Man, Prometheus-like, is at last acquiring the power to control his own genetic future.

To carry on the eugenical work pioneered by Galton, a Eugenics Society is in existence at the present day. Its members, predominantly academics and scientists, hope collectively to influence government policies through normal pressure-group activities. But, however seriously they take eugenics, probably few of them see themselves as having a mission to save civilization from imminent collapse, or seriously expect that eugenics will shortly replace the programmes and ideologies of the existing political parties; nor would they present eugenics as a science of man that was making redundant all previous speculations in philosophy, history, and sociology. These, however, were precisely the aims and ambitions of those who formed the original 'Eugenics Education Society' in the winter of 1907–8.

'The present writer believes that eugenics is going to save the world', wrote a leading spokesman, Caleb Williams Saleeby, in 1909,[2] and these words capture the mood of exuberant optimism within a movement which seemed about to sweep aside all the obstacles in its path. A similar state of affairs existed in America; one historian has calculated that on the eve of the First World War the general journals were carrying 'more articles on eugenics than on the three questions of slums, tenements and living standards, combined'.[3] Yet, whereas the American eugenics movement has been described and analysed in an excellent book by Mark Haller,[4] its British counterpart has received astonishingly little attention. Indeed, it requires something of an imaginative effort to realize that earlier in the century eugenics was by no means a peripheral concern, appealing to a small coterie of enthusiasts and cranks, but an important challenge to politicians and academic theorists alike.

The reasons for the rapid development of a eugenics movement and the forms which it assumed constitute the subject matter of this monograph. The topic has a two-fold interest and importance. On one level, it can be seen as a rather unusual instance of interaction between scientific investigation and political speculation. There have, of course, been many attempts to remove the accidental and the random from political life and to convert government into a science, whose recognized and objectively valid truths could be enforced by the appropriately trained experts. Yet most attempts to construct a science of politics do little more than assert that politics should be made *analogous* to the operations of the physical sciences. Eugenics, however, purports to be nothing less than the *direct application* of the laws of physical science. The plausibility of this claim has been enhanced by distinguished biologists placing their professional reputations and achievements at its disposal; this personal involvement by a group of scientists in a proselytizing campaign is also sufficiently unusual in itself to merit further examination.

Eugenics must also be understood in the light of the problems being experienced by the various industrialized societies in which it took root. These problems varied, but in Britain the eugenics movement undoubtedly owed its initial impetus to the growing realization of the contraction of national power; operating outside the parameters of party politics, eugenists secured a hearing by providing a startling analysis of this relative loss of power, while at the same time holding out the promise of national regeneration to those who heeded their message. Like other responses to this situation—the tariff reform campaign, the National Service League, and the search for greater National Efficiency, for example—eugenics can also be seen as symptomatic of the emergence of a new 'Radical Right' in British politics. Here is a large field of investigation to which historians are only slowly turning their attention, but a better understanding of the place of eugenics in British politics before 1914 may assist its elucidation.

1

Intellectual Origins

'The followers of Calvin and John Knox should all be Eugenists', James Barr once declared, 'because the teaching of those great men on preordination and predestination fits in exactly with modern views on inheritance from the germ plasm'.[1] In their anxiety to present eugenics as a respectable creed, of which no one need be afraid, eugenists were apt to go in for such claims. A bewildering variety of famous religious leaders, politicians and artists from the past were presented as eugenists who had had the misfortune to be born before their time. Even Christ's words about men being content to become 'eunuchs for the kingdom of Heaven's sake' could be construed as advice to the 'unfit' to refrain from parenthood.[2] If all this seems a little silly, it remains nevertheless true that, although the word was not used until 1883, with the publication of Galton's *Inheritance of Human Faculties*, the cluster of ideas involved in 'eugenics' goes back as far as the surviving record of human society. From the moment that man first began to reflect about his destiny, he must have toyed with the idea of improving the human race by arranging that the 'best' types should marry among themselves and produce large families. A hostile critic, G. K. Chesterton, called eugenics 'one of the most ancient follies of the earth'; 'one after another all men with active minds, from the old Greek philosophers to Mr Shaw and Mr Wells, have thought of the notion, looked at the notion, and, in consequence chucked the notion'.[3]

In fact, not all of them had chucked the notion. Utopian speculators have always been interested in schemes for controlling sexual reproduction. The philosopher and eugenist, F. C. S. Schiller, reflected in 1926: 'I cannot quite remember whether I was a eugenist before I read Plato's *Republic*, ever so many years ago; but I have been a convinced eugenist ever since . . .'[4] A reading of Campanella's *City of the Sun*, or More's *Utopia*, must also have set off in many people's minds a train of speculation which later prepared them for acceptance of the eugenics creed.

This makes the vexed question of 'influence' even harder than usual to determine. A writer who was active in the eugenics movement in the 1920s, C. W. Armstrong, recalls how, when only nineteen years of age, he had written a novel, *The Yorl of the Northmen*, describing what was, in effect, a eugenic community, yet at that time he 'had never heard of Sir Francis Galton or eugenics'.[5] Armstrong was probably not alone in this respect. That most interesting piece of journalistic sensationalism, Arnold White's *The Problems of a Great City* (1886), contains, in crude form, a great deal of what later became known as 'negative eugenics'; but had White yet read Galton at first hand, had he picked up his ideas indirectly, or had he reached his own conclusions by means of personal experience and the utilisation of biological concepts with which any literate person would have been familiar in the 1880s?

In a sense, eugenics in its modern form originates with Charles Darwin.[6] The notion that man is part of the world of organic life and subject to the same natural processes of evolution and decay was essential to the development of the new subject. The hypothesis of natural selection was also bound to raise in many people's minds the possibility that, as rational beings, men might learn to control their own evolution; in place of the blind processes of natural selection a deliberate effort might be made to improve the species by attending in a scientific way to the production of offspring. Darwin himself, when presented with this possibility by his cousin, Galton, commented that the object proposed was a 'grand one', but he doubted whether people could ever be persuaded to co-operate intelligently in the matter.[7] However, in Chapter V of that profoundly ambiguous book, *The Descent of Man* (1871), he looked sympathetically at many of the ideas already promulgated by Galton, though he stopped short of positively endorsing them. It is a matter of conjecture whether, had he lived for another generation, Charles Darwin would have come out as a supporter of eugenics, as many of his children did.[8]

In the early 1870s, Galton shared a good deal of Darwin's scepticism about the feasibility of devising an actual scheme of race-improvement. He did not wish to run too far ahead of public opinion, or to shock too violently the accepted social prejudices of his day. Moreover, until much more was known about the mechanism of inheritance in man, it would have been premature to conduct, or even suggest, any social experiments. A warning of what should be avoided was provided by the American, John Humphrey Noyes and his Oneida Community. Starting life as a free church minister, Noyes was led by his theories of 'perfectionism' to break with the orthodox Christian world and set up his own communistic society, whose members were not to be bound by the conventions of the monogamic marriage. 'By a rather harsh discipline Noyes enforced on his followers birth control and stirpiculture (i.e. selective breeding). The number of children to be born each year was predetermined and their parents were selected so as to produce the best possible offspring'.[9] By the time the Oneida Community was dissolved in 1879, fifty-eight children had been born in this way, at least nine of them fathered by Noyes himself.[10] Significantly, the start of this little experiment goes back to the 'Battleaxe Letter' of 1837, in which Noyes's views on sex and procreation were first made public—that is to say, well before the appearance of the work of Darwin and Galton—and the original impulse was a religious-millenarian one. But Noyes later borrowed from Galton in his attempt to make his breeding regulations scientific, as in his book, *Scientific Propagation* (1873).[11] Galton himself, of course, did nothing to countenance the Oneida Community.[12]

Another embarrassment to sober men like Galton was the American, Victoria Woodhull-Martin, an engaging charlatan, whose bizarre career took her through three marriages, numerous liaisons, several well-publicized scandals, an attempt to stand as President of the United States, and advocacy of spiritualism, elixirs of life, communism, sex equality, free love—and stirpiculture. But, despite her invocations of 'science', Mrs Woodhull-Martin had no authority to pontificate on matters of human inheritance, and many of her observations on this subject were ill-informed and nonsensical.[13] It was the backing of responsible and established scientific men which was essential to the progress of eugenics. And this had to wait until such time as biologists had acquired an understanding of heredity that would enable them to explain how parents transmitted certain of the physical and intellectual qualities to their offspring. Not until the Edwardian period had the scientific groundwork been sufficiently well laid for eugenics to become a plausible political creed. Three scientific theories were to prove of especial importance.

Firstly, there was the theory of the German biologist, August Weismann, concerning the continuity of the germ plasm: a theory announced in the late 1880s and quickly accepted as the basis for future investigations. Put simply, Weismann argued that there was a clear distinction between the germ cells, which controlled reproduction, and the body or somatic cells. The germ cells or germ plasm were independent of the somatic cells, and so could not be affected by any modification of the bodily organs caused by use or disuse, by disease or injury. Acceptance of this theory logically entailed a rejection of Lamarck's belief that acquired characteristics could be inherited. This was important to eugenists, who drew from it the further conclusion that environmental reforms could only have a very limited effect on individual human beings, who were the kind of people that they were by virtue of their germ plasm, which the environment could neither change nor improve. One could admit that a bad environment might prevent a child from developing his full potential, just as a poor quality soil would stunt the growth of a plant and prevent it from reaching its normal height; but an environment could not bring out qualities, mental or physical, that were not innately present.

Moreover, however beneficial might be the effects over one generation, of education, public health measures, factory legislation and so on, their beneficial effects could not be transmitted to the next. The process would have to begin all over again. As many political writers and commentators protested at the time, this did not greatly weaken the arguments for social reform nor did it strike at the roots of social progress. But Herbert Spencer, for one, believed that his lifetime's work was being undermined. And, from their quite different perspective, eugenists drew the moral that true progress could only be achieved through racial progress; the level of intelligence, health, energy or beauty could only be raised by breeding from the best stocks and controlling the fertility of the worst.

A theory in some ways similar to Weismann's had earlier been advanced by Galton himself. But Galton's main contribution to the scientific study of heredity came when, taking shelter in the grounds of Naworth Castle to escape a temporary shower, the idea flashed into his mind that 'the laws of Heredity were solely [*sic*] concerned with deviations expressed in statistical units'.[14] Working out the full implications of this approach, Galton virtually created the new science of biometry: that is, the application to biological phenomena, including human beings, of precise and sophisticated statistical techniques. Galton was able in this way to tackle two problems which were still puzzling his contemporaries: firstly, of how to measure accurately the variations between close relatives, or, what amounts to the same thing, the intensity of resemblance between them; and, secondly, of how to decide whether two or more sets of data were causally related or independent variables. Out of these investigations came the co-efficient of correlation and the rapid development of biometry under the direction of Galton's pupil, friend, and future biographer, Karl Pearson. To Galton, eugenics always meant applied biometry. As late as 1907, he could give a public lecture, entitled 'Probability—the Foundation of Eugenics', in which, after discussing his eugenic ideals, he took his audience through a condensed course of statistics. But it was not to be.[15] When eugenics made headway in Britain and America, it derived its main impetus less from biometry than from the exciting advances being made in Mendelian genetics, a subject whose importance Galton never really appreciated.

Thus, the third important scientific break-through which contributed to eugenics was the rediscovery, in 1899, of the Abbé Mendel's famous paper which, although first read in 1865, had subsequently been lost to sight. This paper, however, supplied one of the keys for unlocking the genetic structure of human life, as was quickly appreciated by De Vries and by a group of biologists in Britain, whose acknowledged leader was William Bateson, appointed in 1908 as first Professor of Genetics at Cambridge University. Although Bateson and his fellow workers were largely concerned with breeding experiments on those plants and animals whose comparatively simple genetic structure made them suitable for these pioneering Mendelian investigations (varieties of sweet pea, wheat, and poultry, for example), it was recognized from the start that certain physical traits in human beings also observed the simple laws of gametic segregation which Mendel had analysed in sweet peas. Most human traits, it is true, are not analyzable in quite this way, since they are controlled by more than one pair of genes, but this does not invalidate the Mendelian hypothesis. Such complexities were soon acknowledged by geneticists, who nevertheless felt, and with good reason, that it was now only a matter of time before the mechanism of inheritance in human beings was almost as well understood as it was in the case of the Andalusian fowl. After all, the first ten years of Mendelian experiments had contributed more to an understanding of genetics than a previous century of biological research.

Moreover, the geneticists were not too immersed in their detailed investigations to speculate about the wider significance of their work. Might not the fuller understanding of heredity now being achieved enable men to eradicate scientifically evils and suffering that had baffled philanthropists and reformers for generations? Would it not soon be possible to 'breed out' certain grave hereditary ailments in the way that Mendelian geneticists had learned to breed 'rustiness' out of wheat, and perhaps also to develop mental or physical qualities in men that were generally regarded as desirable? No wonder, then, that genetics and eugenics rushed into each other's arms. Eugenics, it seemed, would advance by the practical application of the knowledge of heredity which genetics was making available. Eugenics would then stand to genetics in rather the same relationship that engineering does to mathematics.

2

The Development of a Eugenics Movement in Britain

Galton, it has been observed, did not feel much enthusiasm for 'Mendelism', and this was to create the most serious problem for the eugenics movement in the 1907–14 period. But at least he saw, around the turn of the century, that enough progress had been made in the scientific field to promote eugenics without fear of exciting universal ridicule. He also correctly sensed that there was a considerable sector of the community that would respond sympathetically to such an initiative. The Boer War panic about possible physical deterioration, the preoccupation with 'National Efficiency', and despondency about the apparent failure of 'environmental' social policies, had between them created a political atmosphere highly congenial to eugenics. In the very midst of this political phase, in October 1901, Galton, invited to give the Huxley Lecture at the Anthropological Institute, chose as the subject of his address, 'The Possible Improvement of the Human Breed under the existing conditions of Law and Sentiment'.[1] Galton was sufficiently encouraged by the reception of his address to advance one step further. At a carefully stage-managed meeting of the Sociological Society on 16 May 1904, with Karl Pearson in the chair, Galton unfolded his eugenic message before a very distinguished audience.[2]

From then onwards eugenics quickly developed from being either an intellectual exercise or a pious aspiration into a political movement, with a variety of institutions of its own for undertaking research and influencing public opinion. At a second appearance before the Sociological Society in February 1905, Galton was able to announce the appointment of Mr Edgar Schuster as a Francis Galton

Research Fellow in Eugenics, a post endowed out of his own resources. In 1907, the Eugenics Records Office became the Eugenics Laboratory under Pearson's direction. When Galton died in 1911, his will provided for the endowment of a chair in eugenics at London University; Pearson was designated as its first incumbent.[3] Already Pearson had achieved substantial progress with research projects sponsored by the London University Biometrics Laboratory, and the more specialized research papers of his workers appeared in the scientific journal, *Biometrika*, although there were other outlets for articles of a popular or of a proselytizing character.

But into the latter field a rival organization entered in the winter of 1907–8: the Eugenics Education Society [hereafter E.E.S.]. It originated as a break-away from a somewhat nebulous organization called the Moral Education League. On 15 November 1907, a meeting took place at the Caxton Hall, attended by certain members of the committee of the latter body as well as by people interested in creating an organization for the furtherance of eugenic ideas.[4] A provisional committee was set up to draft a constitution for the new society. The name, Eugenics Education Society, was adopted, although, according to the *Birmingham Daily Post*, many members at Caxton Hall disliked it, 'and put in a plea for a more readily understood description'.[5] Certain former members of the Moral Education League announced their intention of continuing as a separate body; many others, however, joined the E.E.S., giving it in its early years a large feminine element, which somewhat diminished over the succeeding period. An American lecturer in the social sciences in the London University Extension Courses, Dr J. W. Slaughter, author of *The Adolescent*, was elected provisional chairman. On 9 December the first General Meeting took place, Slaughter was confirmed as Chairman, Mrs Gotto as Secretary, and the newly elected Council settled down to the task of defining aims, preparing leaflets for the information of the public, and setting up a number of sub-committees.[6] By 14 February 1908, the Society was able to call its first A.G.M. at Denison House, which about two hundred 'people of influence' attended, according to the *Pall Mall Gazette*.[7] The Society at once managed to achieve a great deal of favourable publicity by taking up, at the suggestion of one of its most active members, Dr Caleb Saleeby, the issue of inebriate women whom the L.C.C. were about to discharge on to the London streets from an Inebriate's Home. A deputation was organized to lobby the L.C.C., and later an opportunity arose of submitting evidence before a Home Office committee of enquiry.[8] The Society's case was fully presented in *The Times*, and, in short, the Council could be pleased at having got the Society off to a good start.

By the time of the second A.G.M., held in March 1909, the E.E.S. was firmly established. Society rules were adopted which conferred authority on an Executive consisting of the officers of the Society and its sub-committee chairmen, and on a Council composed of its honorary officers and not more than thirty ordinary members, elected annually.[9] By 1910 the Society was sponsoring its own journal, the *Eugenics Review*, and in 1911 preparations were well advanced for an International Eugenics Conference, due to be held in London the following year. By 1914, membership had risen to 634, and there were affiliated branches in Belfast, Birmingham, Haslemere, Liverpool, and Oxford, as well as other discussion groups, as at Brighton.[10]

The achievements of the Society invited respect; it had been asked to give evidence, not only to the Home Office Inebriates Enquiry, but also before the Royal Commission on Divorce, and it had played a significant part in getting the government to set up a Royal Commission of Enquiry into Syphilis and to give legislative effect to the recommendations of the Royal Commission on the Care and Control of the Feeble-Minded. A special Education Conference in 1913 was fully booked in advance, with over 400 teachers and headmasters and head-mistresses present.[11] The Society was especially adroit in getting its point of view placed before public health and medical gatherings. The number of lectures and addresses sponsored by the E.E.S. before 1914 would run into hundreds.

More important than the size of the Society's following, however, was its intellectual calibre and the social prestige of its membership. Given the nature of the Society's work, and the hostility it aroused in some quarters, and the giggling embarrassment in others, it was essential that the organization attract men of weight, gravity, and established reputation. By and large, it did so. Almost the entire biological establishment joined the E.E.S., and many of the most distinguished geneticists took an active part in its day to day work. The

Darwin family were present in numbers; in 1911 Charles Darwin's youngest son, Major Leonard Darwin, consented to become President of the Society, and when the first Galton Day celebrations took place, it was another of Darwin's sons, Francis, a botanist, who gave the Galton Lecture. Of the famous British geneticists, Professor Darbishire and R. M. Lock were both members; so too were the zoologists, Professor J. A. Thomson and Professor Punnett. The one notable absentee was William Bateson, whose cantankerous behaviour caused the eugenics movement some embarrassment. In fact, Bateson's criticisms of eugenics, such as they were, did not strike at the fundamentals of the creed, and his attitude of hostility may have sprung mainly from personal distrust of the leaders of the E.E.S., or perhaps simply from native cussedness.[12] In the event it was possible for Bateson to be quoted approvingly both by eugenists and by the enemies of eugenics. Consideration of his case does at least make one realize the damage that the E.E.S. would have suffered if other eminent geneticists had followed Bateson's lead, instead of giving exemplary support and encouragement to its work.

Medical men were second only to biologists in their attachment to the eugenics cause, even if leading eugenists did often express disappointment that so many doctors hung back or were publicly critical. The medical profession, in fact, was divided about the merits of eugenics. The *British Medical Journal*, for example, was openly hostile.[13] On the other hand, eugenists achieved a notable coup at the Congress of National Health, held in Dublin in 1911, where a resolution was passed supporting eugenic principles.[14] It is also significant that one of the Society's Vice-Presidents was Sir James Barr, the President of the British Medical Association in 1913. Also, for a brief interval, the Presidency was held by James Crichton-Browne, the Lord Chancellor's Visitor in Lunacy and author of numerous works on mental and nervous diseases. Pathologists and experts in mental deficiency and abnormality, including Dr A. F. Tredgold, Dr Clouston and Dr Mott, all lectured and wrote frequently on eugenical subjects.

At one time the infant British sociology movement seemed about to be taken over entirely by eugenists, who even after 1908 were regular contributors to the *Sociological Review*.[15] From the ranks of psychologists came Cyril Burt and William McDougall; well-known literary figures like Lowes Dickinson, T. C. Horsfall, and Havelock Ellis supported the new creed. That eugenics was perfectly compatible with the Christian faith was attested by the participation of a number of ministers of religion, most notably Dean Inge, while the famous Congregationalist preacher, the Rev. R. J. Campbell, consented for a time to serve as an Honorary Vice-President. The subject of eugenics was brought before the Church Congress during a lengthy discussion of Heredity and Social Responsibility in October 1910.[16] Another person in holy orders to give the E.E.S. a badge of respectability was the Hon. Edward Lyttelton, Headmaster of Eton. Roman Catholics, on the other hand, were united in bitter opposition to eugenics from the very start.

Politicians tended to keep their distance from the new organization, but Balfour agreed to become an Honorary Vice-President in 1913 (Asquith politely declined), the Liberal M.P., Walter Rea, was a member throughout the period under discussion, Joynson-Hicks joined the Society, and active in the Birmingham branch was the still largely unknown Neville Chamberlain.

It is also interesting to note how many men and women who achieved public fame between the wars first cut their political teeth, as it were, on eugenics. In November 1913, the Oxford University Union carried by 105 votes to 66 a motion that 'this house approves of the principles of eugenics' after an able speech had been made in its support by J. B. S. Haldane, of New College.[17] At Cambridge University a group of undergraduates approached Professor Inge, Professor Punnett, and Mr Whetham with the proposal that a Cambridge University Eugenics Society be formed, and this was done in 1911. The Cambridge body did not choose to affiliate to the E.E.S., but modelled itself quite closely on the London organization. Among its officers one notes the Treasurer, J. M. Keynes, who, in fact, took a life-long interest in eugenics.[18] It may also not be generally known that Harold Laski was a zealous eugenist in the years before 1914; he wrote an article for the *Westminster Review* in 1910, which brought the young man an invitation to take tea with Galton, and he undertook some of the statistical work for the Eugenics Laboratory publication of 1913, *On the Correlation of Fertility with Social Value.* Laski's biographer has dealt with this phase in his subject's career, and shown that it was far more than a transient adolescent hobby.[19]

And yet, despite the involvement in eugenics activities of so many people of intellectual ability—outside London the movement was largely centred upon the universities—the spokesmen for the E.E.S. were haunted by the fear that 'undesirable types' might be attracted. To quote from Major Darwin's Presidential Address of June 1913, 'Eugenics is always in some danger of being used as a dumping-ground for cranks'.[20] The problem was accentuated by the fact that there was nothing to prevent anyone, whatever his views, from *calling* himself a eugenist. George Bernard Shaw's occasional assumption of this title when giving vent to highly controversial proposals, like equality of incomes, and his association of eugenics with the use of the lethal chamber and free love, as in his scandalous address to the Society in March 1910, drove the officers of the E.E.S. wild with rage, but there was little they could do to mitigate the damage.[21]

Initially, there was some genuine confusion in the minds of people who were attracted towards the eugenics movement without fully understanding what it was all about. The idea died hard that eugenics covered any measure that might improve the health and happiness of babies. Others, who knew better, were easily side-tracked, or, in some cases, deliberately posed as eugenists in order to attract attention to some other cause they wished to promote, temperance, sex education, control of venereal disease, the establishment of milk depots, compulsory military service, and what not.[22] Reviewing the course of the 1912 International Eugenics Congress, Major Darwin correctly noted: 'as eugenists we must be careful lest the result be that the field we ourselves have promised to cultivate should go unlaboured. Looking back at our discussions during this Congress, is it not true that in proportion as speakers have had in their minds the attainment of immediate benefits—in proportion, that is, as they have added other aims to those purely eugenic—in that proportion have our debates been animated. And herein lies our danger . . .'[23] The social reformer, as Darwin was aware, sometimes sought the eugenist's support 'in connection with some minor eugenical advantages resulting from his proposals . . . when once the eugenic blessing has been received, all thoughts of hereditary influences are likely to disappear from his mind'.[24] This seems a fair description of Sidney Webb's attempts to palm off the Poor Law Minority Report as a 'eugenic' document, in a tactical move presumably aimed at broadening the basis of his support.[25] It was not, Darwin hastened to add, that eugenists had any quarrel to pick with social reformers as such; but *somebody* had to represent the interests of the unborn, and what other pressure group would perform that task, if the E.E.S. allowed itself to be side-tracked?[26]

Misrepresentation from opponents and from journalists looking for sensational copy was an additional hazard. Effusive silliness from leading E.E.S. spokesmen—Slaughter and Saleeby were sometimes culprits in this respect—often received maximum publicity in the newspapers, while sober presentations of the eugenics case tended to be ignored. It was also unfortunate that the Society took a little time to acquire officers of the desired calibre. For the first few months of its existence it was being effectively run by its young American chairman, Dr Slaughter, and by Caleb Saleeby. Feeling little confidence in these gentlemen, Galton at first hesitated about being associated with the Society.[27] When, in July 1908, he did consent to become a member, the Council of the Society, delighted, promptly elected him as their Honorary President.

This coup, however, led to certain unforeseen difficulties. In late February, Sir James Crichton-Browne, after some hesitation, had expressed his readiness to serve as the Society's first President. Perhaps he felt that the honour given to Galton took from the dignity of his own position. Whatever his motives, Crichton-Browne did very little on the Society's behalf in 1908, and in March 1909 suddenly announced his inability, because of 'pressure of business', to give the Presidential Address at the A.G.M., due to be held in a week's time.[28] The task was taken on by a lawyer, Montague Crackanthorpe, an old friend of Galton's, who went on to serve as President from 1909 to 1911.[29] All this while Slaughter continued to be Chairman.

Then in early 1911 Galton died. The Council initially hoped to appoint another Honorary President. Edward Lyttelton sounded out Lord Rosebery, who wrote a letter of refusal. Arnold White had earlier spoken of asking the Duke of Bedford to serve, in the event of Rosebery declining.[30] In the end, the office was left vacant.

Meanwhile, changes more important to the future of the Society were in the process of being made in early 1911. Firstly, Crackanthorpe resigned his Presidency and suggested as his successor Major Leonard Darwin. The change-over was effected in March, probably to the benefit of the Society, since, despite his services at a critical juncture in its history, Crackanthorpe's blatantly pro-Conservative bias always threatened to damage the eugenics cause. Almost simultaneously, Dr Slaughter announced his unwillingness to act any longer as Chairman.[31] Slaughter had always been a controversial figure. His often vapid and bumptious utterances in newspapers had much annoyed Galton and Pearson. For some time before 1911, apparently under a cloud, he had taken a less prominent part in the Society's work; in 1912 he ceased his lecturing for London University, departed for Latin America, and never returned.[32] With Slaughter's departure, the Chairmanship was abolished. Major Darwin, as President, therefore assumed unchallenged the role of eugenics spokesman, a role which he filled with great skill and dignity, so that, when he died in 1943, *The Times's* obituary writer justly remarked that Leonard Darwin was sometimes 'described as the founder of sane views on population and society, just as his father was regarded as the founder of modern biology'.[33]

But, if the problem of filling its offices with suitable people had been satisfactorily solved by the spring of 1911, there was no cessation to the running feud that had developed within the eugenics movement between the Laboratory and the Society, a feud highly damaging to all concerned. The trouble originated in the rival theories of heredity being propounded by the biometricians and the Mendelians, which, even before eugenics had become a serious movement, had produced a bitter controversy, and implacable enmity, between the main protagonists of the two schools, Pearson and Bateson.[34] That this dispute should divide Laboratory from Society was almost inevitable, considering the fact that Pearson was director of the biometrical investigations being undertaken by the Laboratory, while the Society soon came under Mendelian influence. It is true that Bateson held aloof from *all* the eugenic organizations, but other geneticists joined the E.E.S. and, from quite an early stage, pronouncements by prominent members like Saleeby and Crackanthorpe had lined up the Society on the Mendelian side.

In February 1908, Pearson was sounded out as to whether or not he wished to become President of the E.E.S., but declined, 'saying his work lay more in accurate statistical research and unless the Society intended working on those lines he would rather not be connected with it in any responsible position although ready at any time to give any help he could . . .'[35] In practice, however, Pearson spared no effort to discredit the Society and ridicule its pretensions, and his research assistants followed suit.[36] Only Edgar Schuster became prominently associated with the E.E.S., and found himself as a result in an unpleasant public altercation with Pearson.[37] The luckless Galton was obliged, in 1910, to defend the Eugenics Laboratory against its detractors in the E.E.S., and, had he lived a few months longer, he might have felt obliged to resign from the Society altogether.[38]

After Galton's death the acrimony surrounding the Biometrician/Mendelian controversy grew, if possible, even more pronounced. Saleeby, who was an active temperance worker, was enraged at the results of research done by the Laboratory into the effects of parental alcoholism. Time and again he denounced the biometrical method as futile, virtually accused Pearson's colleagues of 'cooking' their statistics, and hinted that Galton's bequest was being squandered.[39] More seriously, Crackanthorpe, the President of the E.E.S., wrote to *The Times*, dissociating himself from the conclusions of the Memoir on Alcoholism in a letter which also impugned Galton's belief in statistical methods.[40]

Pearson hit back with venom. He had, it is clear, been sorely provoked. One can sympathize with his amazement at the way in which leading members of the E.E.S. lavished obsequious praise on Galton, while denigrating the research techniques upon which Galton had grounded his eugenic message. The crowning irony came in February 1914, when, in the course of the first 'Galton Lecture', organized by the Society, Francis Darwin completely threw over biometrics: 'Galton seems to me like a medieval chemist while Mendel is a modern one . . . it would seem as though the progressive study of heredity must necessarily be on Mendelian lines'.[41] Well might Pearson wonder why the Society bothered to honour Galton at all; 'is it only because a heterogeneous conclave needs some well-known flag to follow, requires the shadow of some great name to sanction its proceedings, regardless of what the flag stands for, or what were the most cherished principles of the man chosen as its fabulous hero?'[42]

Pearson showed great obstinacy in refusing to recognize the important work currently being done by the geneticists, but it should be said, in his defence, that in these pre-war years 'Mendelism' had assumed the status of a new and exciting religion, and some of its acolytes not only conducted themselves with truculent arrogance, but made reckless claims that were bound to excite ridicule outside the ranks of the converted. Thus, both in Britain and America, self-styled followers of Mendel rushed into print with 'proof' that an extraordinary variety of diseases observed simple Mendelian laws, including feeble-mindedness, alcoholism, and neuralgia, and even tried to explain in this way complex human qualities like shyness, laziness and musical ability! Charles Davenport and his workers at the Eugenics Record Office in America were especially active in promoting this kind of spurious research,[43] and David Heron, of the Eugenics Laboratory, had an easy job of exposing the inaccuracies and slipshod nature of their work in his monograph, *Mendelism and the Problem of Mental Defect.*[44] But as a reviewer of Heron's monograph commented: 'Dr Heron and his colleagues have frequently expressed their distaste for controversy, but they have never allowed this reluctance to deter them from expressing their opinions with the utmost frankness. . . . The extraordinary acrimony which characterises the controversial and semi-controversial papers emanating from the Galton Laboratory has long been a matter of comment'.[45] Pearson himself, distinguished scientist though he was, did his cause no good by an apparent eagerness to pick a fight, on the smallest provocation, with any opponent who would take him on.

But Pearson's most outspoken detractor, Caleb Saleeby, was to prove an almost equal embarrassment to the British eugenics movement. Initially, it is true, the E.E.S. owed more, perhaps, to Saleeby than to any other man. The Council minutes confirm that he was at the very heart of all its pioneering activities: drafting the leaflets explaining the Society's aims, serving on all the important subcommittees, and suggesting tactics, like the protest at the closure of the L.C.C.'s Inebriates' Home. For publicity work of this kind, Saleeby undoubtedly had flair. He was a tireless lecturer, and his popular expositions of eugenics appeared in a large number of papers and magazines, of all political complexions. It was also Saleeby who produced the first systematic book outlining the new creed, *Parenthood and Race Culture* (1909).[46]

But in time Saleeby became something of a liability to the Society. His temperance convictions led him to take up a strong stand against alcohol as a 'racial poison', which many other eugenists found unconvincing, and Galton and Pearson were not alone in squirming at the effusive silliness of many of his public pronouncements. Also, not content with proclaiming himself as the original begetter of the phrases, 'racial poison', 'negative and positive eugenics', and 'eugenist', Saleeby was in the habit of claiming to be Galton's favourite son and disciple, a tactic doubly enfuriating to Galton and Pearson, because innocent outsiders sometimes accepted this claim at face value.[47] Finally, members of the E.E.S. seem to have lost patience. At the elections for the new Council held at the A.G.M. in May 1910, Saleeby was not returned. It is not quite clear what had been going on behind the scenes.[48] But it is significant that when in October 1913 Saleeby wrote, offering to give a paper before the E.E.S. on 'The Influence of Nutrition upon Parenthood', the minutes record that the Council decided by eight votes to two that he should not be heard.[49] By this time there was obviously a deep rift, and it is hardly surprising that in his articles and lectures of the immediate pre-war period Saleeby should have broadened his attacks on the 'better dead' school of eugenists to include leading members of the E.E.S. as well as his old enemies, the biometricians. His warmest encomiums were now reserved for Davenport and the research team at Cold Spring Harbor.[50]

Thus the advocates of a creed that was designed to 'save the world' did not attempt to conceal from the wider public the differences within their ranks. Yet although this must have hampered the growth of the movement, it injured its fortunes less than might have been expected.

3

The Issue of 'Racial Degeneration'

There are several explanations for the popularity of eugenics in Edwardian Britain. Its initial growth was undoubtedly assisted by Galton's shrewdness in choosing October 1901 as the date for launching his project of race improvement, for this was a time when the Boer War preoccupation with 'National Efficiency' and the panic about possibly physical deterioration were coming to a climax.[1] Two years later Arthur Newsholme observed: 'The statement that our national physique is degenerating has been so frequently made and so vigorously repeated that if one doubts this fundamental point it is against the weight of public statements made in nearly every journal with confidence and assurance'.[2] The origins of this belief go back at least to the 1880s, if not earlier. It arose in part out of a deep-seated anxiety about whether Britain may not have taken a wholly wrong turning in becoming a predominantly urban, industrial society. Was there not, perhaps, a heavy price to be paid for this abandonment of a way of life more natural, more in tune with the rhythm of the seasons? More specifically, as Gareth Stedman Jones has argued, middle class commentators on the 'social question' from the 1870s onwards viewed with fear the casual labourers and the inhabitants of the slum areas of the big cities; they noted with both disappointment and apprehension that these people had not 'responded' to attempts by legislators and charitable organizations to raise them to a higher material and moral plane, and some were tempted to explain this by the hypothesis of urban degeneration, by a reversed natural selection that was throwing up a biologically distinct sub-species congenitally incapable of conforming to accepted social norms.[3]

urban degeneration

3 *The Issue of 'Racial Degeneration'*

The illusion that human distress was *increasing* was largely created by the greater sophistication of social investigators, who by their graphic descriptions of urban squalor and destitution, and still more, by their presentation of social problems in precise quantitative terms, made these problems seem infinitely graver than had previously been supposed.[4] The social surveys of Charles Booth and Rowntree, it is a commonplace to say, stimulated many middle class consciences, and provided an important spur to social work and to the politics of social welfare. But on others they had rather the opposite effect: inducing despair and casting doubt on the efficacy of environmental reforms.

This continued to be the case throughout the first decades of the twentieth century. An interesting illustration of this is the use made by eugenists of school medical inspection. Initially welcomed by M.P.s like Balfour as a kind of anthropometric survey, which would produce valuable scientific data about the national physique,[5] the inspectors' reports revealed that the vast majority of elementary schoolchildren were suffering from medical defects of one kind or another. Such evidence was eagerly cited by eugenists, who went on to argue that the national physique was deteriorating:[6] a quite unjustified inference, since, prior to the start of school medical inspection in 1908, there were no statistics about the health of children with which comparisons could be made. Yet even Galton, in a letter to *The Times* of 18 June 1909, can be found contending that national degeneration was taking place, as shown by 'results of inquiries into the teeth, hearing, eyesight, and malformations of children in Board Schools'.

In the same letter Galton gave as additional confirmation of his pessimistic assessment of the national physique 'the apparently continuous increase of insanity and feeble-mindedness'.[7] To support this assertion official statistics did exist. They indicated that the numbers of the certified insane had grown from 2·2 per thousand in 1872 to 3·2 per thousand in 1909, and that within the same period the number of insane paupers had jumped up by no less than 130 per cent.[8] In the County of London, where the total population had remained almost static, registered insanity practically doubled in the twenty years prior to 1912.[9]

But there were hazards in attempting to draw general deductions from these figures. Additional facilities for the mentally handicapped, better methods of identifying people in need of institutional treatment, and fuller statistical records, all gave an impression of deterioration which may or may not have reflected the realities of the situation. There are similar problems about using national expenditure on poor relief as a means of quantifying the extent of destitution. These difficulties, however, did not prevent alarmists from quoting these statistics as evidence of a growth in pauperism and 'dependence'.[10]

One might have supposed that the Registrar-General's Reports showing a steady decline in mortality would have done something to modify this picture of a catastrophic decline in the nation's health. But those who held this belief, most eugenists included, drew ammunition even from these reports. The drop in infant mortality which began in the early twentieth century raised forebodings about the greater number of the 'unfit' who were being artificially kept alive by the miracles of modern medicine, and the falling tuberculosis death-rate was also viewed by many eugenists as being a far from unmixed blessing. On the other hand, eugenists could point to an increase in deaths from what Dr Haycraft called 'constitutional diseases', which included diabetes and cancer, as an illustration of the 'racial deterioration' that had already materialized 'as a sequence to that care for the individual which has characterised the efforts of modern society'.[11] The growth of fatalities from cancer was also made much of by Arnold White in *Efficiency and Empire*.[12] Cancer, said another alarmist, was 'almost invariably the outcome of impaired vitality, resulting in constitutional degeneration'; it was also, in his opinion, a hereditary ailment.[13]

But quite the most important of all the statistical evidence used to bolster up the theory of 'national degeneration' came in the form of the reports of the Inspector-General of Recruiting at the time of the Boer War. Arnold White was perhaps the first journalist to seize on to the startling fact that three out of five men presenting themselves for enlistment in Manchester in 1899 had had to be rejected as physically unfit.[14] An anonymous article (penned by General Maurice) in the *Contemporary Review* in January 1902 made this example of the physical unfitness of the industrial working class a major political issue, and set off an agonized debate about the sources of imperial greatness.[15] For how could the British Empire survive if there were not an 'imperial race' to sustain it? Could reliance be placed on the stunted, rickety, disease-ridden wretches who inhabited the slum districts of the large urban centres? The Inspector-General of Recruiting himself commented in his Report for 1902: 'the one subject which causes anxiety in the future, as regards recruiting, is the gradual deterioration of the physique of the working classes, from which the bulk of the recruits must always be drawn'.[16]

But informed opinion did not accept that there was any proof of actual deterioration. The Director-General of the Army Medical Service questioned this theory; so did the Royal College of Surgeons, and so did an Inter-Departmental Committee set up to examine the whole problem at a time when the scare of 'racial decay' was at its height. All these authorities made the obvious point that under a voluntary system of recruiting the quality of the men who present themselves for enlistment will be regulated by the prevailing conditions in the labour market, and thus no deductions can be drawn from recruitment statistics about the health of the population as a whole. Quite apart from this consideration, the Royal College of Surgeons noted that the army rejection rate over the previous ten years scarcely supported the claims of the alarmists: the number of men rejected for failing to meet the standard chest-measurements and for 'imperfect constitution and debility', for example, had actually declined, while the rejection ratio had risen most steeply for loss or decay of teeth. On the other side, it could be argued that the minimum height and weight laid down for army recruits had been successively lowered in the latter half of the nineteenth century, and that this alone explained the statistics cited above. Less convincing were comparisons between the height and weight of the average twenty-year-old recruit in 1900 and the height and weight of the average youth of nineteen years as measured by the Anthropometric Committee of the British Association in the 1880s, since here like was not being compared with like. In the absence of compulsory military service, or a universal system of school medical inspection, or of a national anthropometric survey, sustained over a period of several decades, no-one could say with any certainty whether the national physique was degenerating or improving.[17]

All that the evidence of the recruiting officers' reports showed was that a high proportion of the working class population was in poor shape. In the view of the Inter-Departmental Committee, this was mainly due to a bad environment, inadequate and unsuitable nutrition, and the prevalence of various practices injurious to health. Like other inquirers into the alleged physical deterioration of the race, the Committee concluded its labours with a battery of reform proposals, such as public subsidies for a school meals service, state encouragement to organizations providing physical training, an extension of town planning by local authorities to prevent overcrowding, action to reduce smoke pollution and to combat adulteration of food, instruction of the poor in domestic science and child rearing, and so on.[18] The environmentalists had achieved a considerable victory, and the Report of the Inter-Departmental Committee was to provide valuable ammunition for the Radical social reformers in the years up to 1914.[19] The more intelligent and honest of the eugenists, like Havelock Ellis, admitted that it could not positively be proved that the race had fallen below the level it had occupied seventy years ago, but argued that it was a 'terrible thing' that despite so much expenditure of money and effort it was equally impossible to record any incontrovertible improvement.[20]

Most eugenists anyway stuck to their beliefs in biological deterioration, and claimed that environmental reforms, taken by themselves, were likely to do more harm than good. Some credence was still given to an earlier theory, publicized by James Cantlie in a lecture of 1885, that big cities like London so sapped the strength and vitality of their inhabitants that second-generation Londoners were of a feeble and debilitated type, scarcely capable of reproducing themselves.[21] This version of the theory of urban degeneration still had its adherents in the 1900s. At a British Medical Association Conference in 1905 one speaker recalled Mr Cantlie's challenge to his audience to produce a fourth generation Londoner, and claimed that this challenge had never been answered.[22] Even Havelock Ellis believed that 'the town population is not only disinclined to propagate; it is probably in some measure unfit to propagate'.[23]

But even if this factual assertion could have been substantiated, eugenists were not, as a body, attached to this particular hypothesis. Firstly, it pointed to the need for a 'Back to the Land' movement, rather than to a policy of race culture. Secondly, the supposition that a particular kind of environment produced progressive physical deterioration, in a biological sense, could be dismissed, as Archdall Reid dismissed it, as a Lamarckian fallacy.[24] It was certainly absurd to complain about the degeneration of the race and then go on to suggest as a remedy compulsory military training![25] Besides, whatever Cantlie may have believed, there was no evidence for town life destroying the breeding capacities of the urban poor. Sidney Webb expressed relief that this was so. 'If the decline in the birth-rate had been due to physical degeneracy, whether brought about by "urbanization" or otherwise, we should not have known how to cope with it', he wrote.[26] In fact, of course, it was the high fertility of the urban poor which was alarming eugenists, not their inability to propagate themselves. More plausible was the hypothesis that city life 'selected' stocks with qualities to which civilized societies attached little value. A low type of humanity, it was sometimes said, could survive and multiply in the modern city, in the same way that maggots thrived and multiplied in putrefying substances; hence, to quote one contributor to the *Eugenics Review*, the development of 'a race of men, small, ill-formed, disease-stricken, hard to kill'.[27] This proposition, however, logically led on, as Sidney Webb was quick to perceive, to demands for drastic changes in the urban environment, demands which were hard to reconcile with a 'hereditarian' position.

Thus, the main theory advanced by eugenists was the one succinctly stated by the Whethams, when they wrote: since 1875 'a wrongly-directed selective birthrate has been established, and the race is threatened with decay'.[28] This brings us to the crux of the case for eugenics: the meaning of the differential birth-rate. The over-all decline in the birth-rate since the 1870s was in itself a cause for alarm to many eugenists, although there were differences of opinion about whether deliberate family limitation should be encouraged or censured. All, however, could agree that it was disturbing that the birth-rate was not falling evenly throughout the community, and that the effects of this were being felt more strongly at the upper than at the lower end of the social scale. But so it was. Professional families were scarcely reproducing themselves, while large families were still common among unskilled labourers and the very poor. An unofficial Birth-Rate Commission, set up in 1913, later confirmed this picture by showing that amongst the upper and middle classes there were approximately 119 births per thousand married males aged under 55 years; the comparable figure for skilled workmen was 153, and for unskilled workmen 213.[29] There were exceptions, such as the extremely low fertility of the working classes in the textile towns, caused by the high proportion of females in gainful employment. But it was broadly true to say that the birth-rate fell as the standards of comfort rose. And this divergence between the average size of working and middle class families constituted, in the Whethams' opinion, 'the most important social phenomenon of the past forty years'.[30] The higher death-rate in working class districts did something, of course, to reduce the difference in *effective fertility*, but a large gap still remained. Moreover, this was apparently a relatively recent demographic trend, since the further investigations were pushed back into the nineteenth century, the smaller the class differentials uncovered, and the latter could be accounted for almost entirely by the later age at which middle class marriages were contracted. 'This result', said the National Birth-Rate Commission, 'seems to be of great and serious significance. For so long as we knew that the ranks were being replenished mainly from below it was possible to hold that this had probably always been the case, and to believe that, as the nation had prospered in the past, so it would probably be fitted to prosper in the future under such conditions. This belief, however, appears to be no longer open to us'.[31] One of Pearson's assistants, David Heron, the organizer of one of the pioneering pieces of research in this field, was categorical in his insistence that 'the relationship between inferior status and high birth-rate has practically doubled during the last fifty years, and it is clear that in London at least the reduction in size of families has begun at the wrong end of the social scale and is increasing in the wrong way'.[32] The result, as Karl Pearson repeatedly proclaimed, was that twenty-five per cent of the population were producing fifty per cent of the next generation. The racial mixture of the British people, it was alleged, was therefore undergoing a rapid transformation, and since the 'worst' stocks in the community were increasing while the 'best' stocks were dying out, the process could signify nothing less than 'national degeneration'.

Even the critics of eugenics had some reason to feel alarm at the fact that those who were best equipped to provide their children with a good home and upbringing were opting out of their responsibilities, while the very poor were attempting to rear large families in an unhealthy and uncongenial social environment.[33] Full blooded eugenists, however, went well beyond this point, and based their arguments, as we have seen Pearson did, on the premise that a rough correspondence existed between social class and differing biological stocks. The racial deterioration scare only made sense if one assumed that the poor were congenitally inferior to their 'social betters'. 'Reform eugenists' pointed out that although the working class population undoubtedly contained more than its fair share of the feeble-minded and other degenerate types, there was 'not a scintilla of proof' for the theory of racial deterioration, and, as Saleeby argued, biometrical investigations into differential fertility were neither here nor there so long as their authors omitted to consider 'the differences in nurture, education and opportunity between these two classes'.[34] But such warnings largely fell on deaf ears. Heron's researches had demonstrated that there was a relationship between inferior social status and a high birth-rate;[35] this became transformed into Karl Pearson's dictum that 'Mr Heron has indeed shown us that the survival of the unfit is a marked characteristic of modern town life'.[36] From there, it was a short step to emotional effusions, such as the following from Dr Tredgold: 'There is not the slightest doubt that the decline [in the birth-rate] is chiefly incident in—indeed, one may say practically confined to—the best and most fit elements of the community, whilst the loafers, the incompetents, the insane and feeble-minded, continue to breed with unabated and unrestrained vigour'.[37]

Possibly the most bitter comments on differential fertility came from the Whethams. Conscious family limitation was an abomination, they said, because it prevented the operation of natural selection. 'Survival of the fit is of no use to the race unless the fit produce and rear a preponderating number of offspring. In the modern civilized life of mankind, at all events, the best chance of survival does not always mean the probability of the greatest number of children'.[38] The

Expansion of the worst elements

Whethams could hardly contain themselves as they surveyed 'the utterly wanton, selfish, and senseless restriction of the birth-rate among the intellectual and able classes of the community'. Victorian England, they argued, had achieved a peak of culture and intellectual achievement previously reached only in the short-lived renaissances of the thirteenth and sixteenth centuries. Yet instead of consolidating and developing these achievements, the intellectual leaders of Western Europe, and of Britain in particular, were ensuring a new era of decadence and decline by committing 'race suicide'; 'Surely the gods themselves must weep over the perverse stupidity of the human race, who three times in a thousand years have held such possibilities of glorious development in their hands, and three times have compassed their own destruction . . .'[39]

Speculating about the hypothetical characters of those 'suppressed personalities', i.e. the babies that had *not* been born since 1875, was an activity in which free play could be given to the imagination! The absence of these unborn babies could be used, for example, to explain any social tensions or political difficulties in which the nation happened to find itself. If there was a lack of men of the first order in all walks of national life, that was due to the differential birth-rate.[40] The shortage of suitable army officers was 'largely a biological phenomenon', caused by the low fertility of the stocks from which the higher military ranks were customarily filled.[41] The Whethams actually went so far as to suggest that Germany's recent superiority over Britain as an industrial and trading community was due to the fact that the German 'birth-rate did not begin to fall systematically till twenty years later than that of Great Britain, and even now has only sunk to the comparatively high figure of 33 per thousand, as against our 26 per thousand . . .'[42]

During the inter-war years evidence emerged which suggested that the disparity in size between working and middle class families was beginning to narrow. But this lay in the future. In the Edwardian period eugenists could still claim that the middle classes, with all their good and bad qualities, would soon follow the bison and dodo to extinction, while the population became increasingly recruited from a 'tabid and wilted stock'.[43] Eugenists enjoyed frightening themselves by projecting into the future what would happen if present trends were allowed to proceed unchecked. George Mudge in the *Mendel Journal* described a 'degenerate stock' then numbering some two hundred individuals 'carrying or manifesting the tubercular, paralytic and epileptic diathesis'. 'If we go forward to the next generation', he wrote, 'there may be two thousand of them, and in the third generation twenty thousand of them! An army of epileptics, paralytics and tuberculates!'[44] The Whethams asked their readers to contemplate the prospect of the whole population of England gradually becoming feeble-minded or (what they apparently regarded as a closely related condition) 'unintelligent casual labourers'.[45]

The assumption behind such wild talk as Mudge's was that degeneracy was hereditary; the editor of the *Eugenics Review* himself claimed that 'all students of the subject' now believed this to be the case.[46] Allied to this belief was the conviction that degeneracy was a *general condition* which manifested itself in many different forms. 'It is very striking, after one has studied a great many pedigrees of unhealthy, weak-minded, and neurotic stock', wrote the Whethams, 'to realise how often alcoholism in the men seems to correspond with a tendency to tubercular disease in the women, and how both are interchangeable with a low or unstable type of mental character. One gets a very strong impression that, in a certain sense, these things are symptoms rather than diseases, and that it is to the stock which produces them rather than to the individual who suffers from them that we should turn our attention'.[47] In his statistical study of the 'insane diathesis', Heron found, or so he claimed, 'an inherited tendency to general degeneracy, which is something wider even than the vague "insanity", where many types are clubbed together under one name. . . . It would appear as if the stock suffered from hereditary determinants which were unstable in character, and that the existence of such unstable determinants may be the mark of "degenerate" stock'.[48] 'Degenerate' was also a category much employed by the Liverpool physician, Dr Rentoul; he used it, he said, because the word was familiar to most readers through the work of Max Nordau, and because it usefully covered not only lunatics, but sexual perverts, neurotics, alcoholics, kleptomaniacs, and other pathological types.[49] To give one final example, Pearson claimed to have accumulated at the Eugenics Laboratory 'endless pedigrees demonstrating how "general degeneracy" runs in stocks, epilepsy, insanity, alcoholism, and mental defect being practically interchangeable, numberless members failing to reach normality'.[50]

American eugenists made a particular hobby of documenting the history of these degenerate families over several generations. R. L. Dugdale set off the trend in his often quoted study of 'the Jukes', who, it was alleged, had in the course of five centuries produced 709 descendants unfit for society. Ironically, Dugdale himself, as Havelock Ellis notes, was concerned to prove the influence of bad environment rather than of bad heredity.[51] But later investigators, using his study as a model, had strong hereditarian sympathies; and their detailed chronicles of the misdeeds, diseases and disabilities of the Kallikaks, the 'Nam Family', the 'Tribe of Ishmael', and so on, were designed to support the theory that there were stocks so 'tainted' that society, in its own self-defence, was entitled to prevent their multiplication.[52]

To describe these 'defective stocks' eugenists employed a number of emotionally charged terms. People with this genetic endowment were 'submen', 'moral perverts', 'low grade types'.[53] It was sometimes suggested that society was becoming differentiated into two almost distinct groups, or even species, like the Eloi and Morlocks in Wells' *Time Machine* (1895). 'On the one hand', wrote Tredgold, 'there are those of sound, unimpaired constitution and vitality who are on the "up-grade" and who are adapting themselves to the demands of the time—the biologically fit. On the other hand, there are those springing from germ-plasm which is so impaired that this adaptation is impossible, who are on the "down-grade" and falling out in the march of civilisation—the biologically unfit'.[54] In 1937, Dr Cattell was to warn that, if present demographic trends continued, the population would eventually split into 'two distinct intelligence groups, as distinct socially as most Indian castes and more distinct biologically than most races'.[55]

However, during the Edwardian decade one observes a slight decrease in references to general degeneracy, and many more references to what was widely regarded as the root cause of a variety of pathological conditions: mental defect. One example of this is the discrediting in eugenical circles of Lombroso and his 'science' of 'criminal anthropology'. Eugenists were from the start attracted to the notion that there existed a category of persons called 'hereditary' or 'inborn' criminals; the respected pathologist, Dr Clouston, could talk early in the century about 'organic lawlessness' being 'transmitted hereditarily'.[56] For those who believed this, support could be derived from a reading of Lombroso, who argued that criminals were a distinct anthropological type, with marked physical peculiarities, like prehensile feet and flattened noses, who could be seen as a degenerate throw-back from an earlier phase of human evolution; they 'bred true to type' and were quite resistant to curative treatment.[57]

Initially, Lombroso was given a respectful hearing by British eugenists. But his reputation as a serious criminologist was fatally undermined by the publication in 1912 of Charles Goring's *The English Convict: A Statistical Study*, which concluded: 'If there is any real association between physical character and crime, this is so microscopic in amount as not to be revealed'. Eugenists were all the more likely to take this conclusion seriously, in that Goring was one of Pearson's fellow-workers at the Eugenics Laboratory. Yet Goring also dismissed environmentalist explanations of crime, and believed that criminality was restricted to '*particular stocks* or sections of the community'. The most important cause of criminal behaviour Goring found to be mental defect; between ten and fifteen per cent of the prison population examined were suffering from this disability.[58]

Havelock Ellis, who had introduced Lombroso to his English audience in his book *The Criminal* (1890), was quick to seize upon this idea. Already in 1911, in *The Problem of Race-Regeneration*, Ellis was trying to estimate the proportion of criminals who were feeble-minded. He quoted an investigation at Pentonville Prison, where, even after prisoners too mentally affected to be fit for prison discipline had been excluded, eighteen per cent of adult prisoners and forty per cent of juvenile offenders were found to be feeble-minded. Another authority ventured the opinion that only four to five per cent of criminals came from parents who were 'really sound'.[59] Groups on the fringes of society who were especially likely to find themselves in prison, like prostitutes and tramps, were also thought to contain a very high proportion of the feeble-minded. On the eve of the war these suppositions were beginning to be seriously examined by psychometricians, armed with Simon-Binet tests, a movement already well under way in the United States.[60] The investigators reached differing conclusions, but agreed that crime and mental defect were significantly correlated.

The whole issue of feeble-mindedness received wide publicity with the publication, in 1908, of the Report of the Royal Commission on the Care and Control of the Feeble-Minded. The Commission accepted Dr Tredgold's estimate that these defectives produced very large families, perhaps 8·4 on average; although some of the offspring were still-born or died in early infancy, the survivors still exceeded the size of the normal family. Moreover, if, as most experts believed, two-thirds of these children were themselves likely to be feeble-minded, the prospects were ominous.[61] The absence of any legal powers to keep the feeble-minded under custodial care made it impossible, until the Mental Deficiency Act was passed in 1913, to do anything to control their fertility. This explains why the campaign for the enactment of restrictive legislation should have absorbed so much of the time and enthusiasm of eugenists in the 1908 to 1913 period, and also the prominence within the movement of experts on mental defect, most notably Dr Tredgold.

Brooding obsessively as they did on the problems of disease and degeneration, British eugenists not surprisingly took on the role of prophets of doom. But this brought them into emotional rapport with many people outside their ranks. Warnings of national disaster, the end of civilization, the prospect of a new 'dark age' filled the newspapers and magazines in the early years of the twentieth century. It was characteristic of the period that Balfour, when giving the Sidgwick Memorial Lecture in 1908, should have taken as his theme 'Decadence', and characteristic of the man that he should have reached no definite conclusions about it.[62] The reasons for the prevailing gloom are many and complex, but it is clear that the aftermath of the Boer War had made many Englishmen uneasily aware of the fragility of their Empire, and dimly conscious that Britain as a world power was on the wane. Comparisons with the collapse of the Roman Empire were common, and so was an obsession with what caused the rise and fall of civilization.[63]

To these perplexing problems eugenics purported to supply an explanation and a resolution. The symptoms of decline observeable in modern Britain should be seen as 'problems of national physiology', so William Bateson, the geneticist, contended.[64] Galton had long believed that the rates at which various social strata contributed to the population at various times provided the main cause of the rise and fall of nations.[65] These sentiments were echoed by Pearson, who declared that 'selection by parentage is the sole effective process known to science by which a race can continuously progress. The rise and fall of nations are in truth summed up in the maintenance or cessation of that process of selection'.[66] The acquired skills of a civilization might for a time 'mask' its biological decay, but in the last analysis the 'racial fitness' of the population was the factor which determined its survival chances. That was why Britain's differential birth-rate, unless quickly checked, was thought by eugenists to foreshadow the collapse of the British Empire.

4

Eugenics, Empire and Race

From another standpoint the eugenics movement can be seen as part of the wider quest for national efficiency, which so dominated British political thinking in the opening years of the twentieth century. Galton himself spoke about encouraging 'a more virile sentiment, based on the desire of promoting the natural gifts and the National Efficiency of future generations'.[1] The phrase appears again in his characterization of the feeble-minded as 'a very serious and growing danger to our national efficiency': a danger which only 'a eugenic victory' could avert.[2] Arnold White came close to identifying the two creeds in his article on 'Eugenics and National Efficiency' in the *Eugenics Review*.[3]

There was another reason why eugenists presented their cause as a 'patriotic' one. Only patriotism, they argued, could give large numbers of men the incentive and the discipline which would enable them to purify their stock and initiate schemes for racial improvement. In the words of the Whethams: 'The power of combination and organization, the social instinct, readiness for self-sacrifice to the common good, love of home, country, and race—in a word, patriotism—all are needed to bring to birth and to develop a nation fit to hold its own in the fiery trial of war, and in the slow, grinding stress of economic competition'.[4] Needless to say, the Whethams assumed that success in the earnest rivalries of peace and war would go to the racially fit.

'The nation which first subjects itself to a rational eugenical discipline is bound to inherit the earth', argued F. C. S. Schiller.[5] Galton, Pearson, James Barr, and many others rang variations on the same theme.[6] Coupled with descriptions of the glorious destiny which awaited Britain should she only adopt eugenics and thus place her statecraft on a scientific basis, were warnings that, should this opportunity be missed, rival powers would gain at her expense. One eugenist made this point in a very specific way; just as the Germans had exploited many British scientific discoveries and applied them with a practical skill lacking in the land of their origin, so, too, might they seize on the possibilities which eugenics offered: 'These, again, scientifically tabulated, are the work of an Englishman, but the German, with his accustomed painstaking capacity, will probably be the first to turn them to advantage'.[7] In the 1930s, some eugenists pointed out, with mingled emotions of admiration and alarm, that this was already happening. Conversely, failure to do anything about the differential birth-rate would, in the view of most eugenists, so impair the vitality of the British population as to lay the Empire open to the depredations of her more virile neighbours.[8]

As the holder of a vast Empire, the British, said the eugenists, had particular grounds for preserving and enhancing the vigour of their stock. Lord Rosebery's vague phrases about the Empire depending upon an efficient imperial race were given a very particular meaning. In Galton's words, 'To no nation is a high human breed more necessary than to our own, for we plant our stock all over the world and lay the foundation of the dispositions and capacities of future millions of the human race'.[9]

Improving the *quality* of the British population—or at least arresting its decline from A1 to C3 status—was something upon which all eugenists could agree. Opinions differed, however, as to whether the trend towards smaller families (leaving aside for the moment its differential aspect) was a healthy or a dangerous development. The issue clearly had implications for Britain's role as a world power. Those who advocated population growth argued that, all other things being equal, the Power with the greater manpower resources would have the advantage in international rivalries, and also both the means and the incentive to promote ambitious colonizing ventures.[10] There was also the argument that an Imperial Power needed to have the self-confidence which would impel it to disseminate its racial type throughout the globe. Pearson could recall 'no case of a race with a very low birth-rate maintaining or creating a position for itself in the assembly of nations'.[11] The authors of an Additional Note to the Birth-Rate Commission also emphasized the imperial issue: 'If we value our national life should we not desire its diffusion? For the sake of the backward types even . . . should we not desire the preservation and expansion of our people?'[12]

Many eugenists who might have been tempted to give an affirmative answer to this question refrained from doing so because they believed that a high birth-rate would lead to 'over-population', which in turned increased the possibility of war. This was one of the favourite theories of Montague Crackanthorpe's *Population and Progress* (1907), and it elicited from Galton the interesting suggestion that some future Hague Conference might consider the 'limitation of populations', and in so doing get at the heart of the German problem.[13] Eugenists were bound to be attracted by any further evidence in support of their contention that the great controversial issues of history and politics were, properly understood, ***biological*** issues.

But the insinuation of Crackanthorpe and all 'reform Eugenists' was that wars were evils which should at all costs be prevented. Not all eugenists shared this view. One recalls the oft-quoted words of Karl Pearson: 'You will see that my view—and I think it may be called the scientific view of a nation—is that of an organized whole, kept up to a high pitch of internal efficiency by insuring that its numbers are substantially recruited from the better stocks, and kept up to a high pitch of external efficiency by contest, chiefly by way of war with inferior races, and with equal races by the struggle for trade-routes and for the sources of raw material and of food supply'.[14] Here is Social Darwinism, red in tooth and claw, of a kind that was also much in vogue in imperialistic circles in Germany.

In Britain, these convictions led to a certain rapprochement between certain eugenists and those campaigning for compulsory military service. Colonel Melville, Professor of Hygiene at the Royal Army Medical College, argued in the pages of the *Eugenics Review* that military service was 'eugenically useful because it [kept] prominently before the community ideals of physical fitness and efficiency as well as of courage and patriotism'. 'It may be', he added, 'that an occasional war is of service by reason of the fact that in times of danger the nation attends to the virility of its citizens'.[15] Predictably enough, we also find Arnold White, representing the National Service League, jumping up at the International Eugenics Conference, to draw attention to 'the eugenic effect of discipline, of training, of obedience, and of learning the secret of willingness to die for a principle'.[16] And Sir James Barr wanted universal military training and a 'cultivation of the military spirit to arrest the decadence of the nation'.[17]

But perhaps a majority of eugenists in Britain (and also in America) took the opposite view. War, they argued, was 'dysgenic', because it led, as one American put it, to 'waste of germ plasm'.[18] Wars may have had beneficial effects in the past. But modern warfare, it was argued, damaged the national physique, because (to quote Carr-Saunders) 'those who are exposed to risk of death in battle are men who have not yet married or who have not completed their families, and who are mentally and physically fitter than the average. The direct effect of war is thus to bring about a loss of births in the families of the fitter men and to kill off a certain proportion of them. It must tend to lower the average mentality and physique of the population'.[19]

That this was the 'official' line of the Eugenics Education Society is suggested by the endorsement given to it by its respected President, Leonard Darwin.[20] But it was an issue on which British eugenists were rather seriously divided. These differences came to the fore at the 1912 International Conference, where the American, Kellogg, gave an address stressing the dysgenic consequences of war, which provoked sharp dissent from Colonel Melville, Arnold White, and Colonel Warden. The Society kept out of the dispute as long as it could.[21] An editorial in the *Eugenics Review* gave approval to universal physical training, whether of a military or of other kinds, but pointed out that this provided no substitute for the more important task of 'breeding from sound and healthy stocks'.[22] At the same time the Society declined to send delegates to a Peace Conference, in 1912, on the grounds that the prevention of war 'did not come within the scope of [the Society]', still less would it have anything to do with the Anti-Conscription League.[23]

The moment for decision finally came in August 1914. But to those who have examined the utterances of eugenists during the preceding years it comes as no surprise to discover that, although the E.E.S. accepted its patriotic obligations and supported the war effort, it did so with considerable foreboding. In an unsigned article in the *Eugenics Review* in October 1914, the fear was expressed that the dysgenic effect of war would be especially felt in Britain because of the voluntary basis of her army. (In fact, even under a system of conscription, the physically debilitated and the mentally sub-normal would have been excluded from service). Characteristically, the Society was also worried at the prospect of widespread suffering among the professional middle classes: to alleviate which the Society participated in the charitable activities of the Professional Classes War Relief Council.[24] The high death-rates among the officer class, the government's adoption of 'war socialism' to offset shortages of essential commodities, and the upsurge of revolutionary socialism in the last phase of the war, came as yet further evidence that modern war threatened most of those values and social traditions that eugenists held dear. Most eugenical pronouncements in the 1920s have a decidedly 'pacifist' flavour.

But, with Karl Pearson a prominent exception, the British Eugenics Movement had never been belligerently patriotic, although it attracted some outspoken and perfervid nationalists to its ranks. The Social Darwinism of Pearson and a few kindred spirits has misled some historians into supposing that there was a close and logical connexion between the eugenist's creed and the glorification of war. This is to overstate the importance of Pearson, who was in many respects not characteristic of British eugenists. It must also be remembered that the E.E.S. contained many members whose views on international relations derived from humanitarian liberal and ethical socialist sources. Even Dean Inge, whom no-one could suspect of liberal or socialist proclivities, can be found protesting, in 1913, against militarists who advocated a high birth-rate because they regarded men as mere 'food for powder'.[25] Nor did all eugenists by any means accept uncritically Pearson's Social Darwinism or similar attempts to apply the notion of a 'struggle for existence' to the sphere of foreign affairs.[26] Finally, even the strongly patriotic contingent was held back from jingoistic excesses by the recollection that eugenics was a world wide movement; there were frequent meetings and a continuous interchange of ideas between themselves and colleagues in other countries, particularly America. And, of course, eugenics claimed to be merely the practical application of scientific truth, and science knows nothing of national frontiers.

The contention that British eugenists were not 'super-nationalists' or glorifiers of war might be countered with the objection that they were nearly all of them pronounced racialists, as their very language proves. Phrases like 'the traditions of the race', 'racial instinct' and 'race-regeneration' occur with monotonous regularity in eugenical literature. This by itself, however, is not conclusive, since these phrases were also regularly employed by contemporaries who cannot by any stretch of the imagination be called 'racialist'. In the early twentieth century the word race seems to have been interchangeable with 'nation', 'community', or even 'people'. Did British eugenists mean anything specific when they invoked the issue of race?

The main interest of eugenists was probably that of determining the racial composition of the inhabitants of the United Kingdom. Such a study of Britain's own 'local races' might, it was hoped, be made an integral part of a national anthropometric survey, so that it could be ascertained what contributions were being made to the national life by Celts, Anglo-Danes, and the like.[27] Karl Pearson once observed that 'the science of Eugenics is in fact only highly developed and *applied* anthropology',[28] and much eugenical discussion is frankly derivative: a mere re-hash of current anthropological views on the main racial groups into which the peoples of the world naturally divide. There is much mention of the 'three races', Teutons, Alpines, and Mediterraneans, and some windy attempts to explain the whole of European history in terms of the interaction of these contending racial groups.[29] But it was changes in the racial balance of Britain's own population which remained the dominant concern, even though this never aroused such fierce emotions as did the protest being made by the American eugenists against the 'swamping' of the native stocks in the United States by the greater fertility of the newer immigrant groups.[30]

In Britain, immigration was a subsidiary issue, but it attracted some attention all the same. Predictably, eugenists were unhappy

T.K

about its probable biological consequences. Tredgold argued that, while the colonies were drawing away healthy and energetic British men and women, a large proportion of immigrants were 'the scourings of Europe, and immeasurably inferior to the British people whose place they fill'.[31] The Whethams emphasised that the colonies all had immigration controls which kept out the mentally defective, the consumptive, and the pauper, whereas in England only the 'more notorious criminals' were excluded. If the dregs of Eastern Europe were not prevented from entering the country, the end product would be 'an Anglo-Slavonic hybrid stock' of dubious racial value.[32]

The Whethams tastefully avoided saying, in direct language, that it was Eastern European *Jews* whom they were anxious to exclude from Britain's shores. But, in fact, what they and many other eugenists were engaged in doing was continuing the agitation against Jewish immigrants which had originated in the 1880s and had driven the Balfour Government into passing the Aliens Act in 1905. It is significant that several of the most outspoken 'restrictionists' around the turn of the century, notably Arnold White, were later drawn into the E.E.S.[33] Also significant is the investigation of Jewish immigrant children which Pearson began just before the First World War in conjunction with Margaret Moul. Using pre-Binet tests, he concluded that Jewish children were of approximately the same intelligence as Gentile children, but that they were inferior in physique and somewhat dirtier—an important factor, he said, should London ever be struck by a great epidemic. The whole investigation rested on methods of enquiry and on premises of dubious validity, but all that need be done here is to describe the practical proposals which Pearson and Moul derived from their findings. 'Let us', they wrote, 'set a standard for immigrants, say 25 per cent higher than the mental and physical averages of the native population—and in the present state of our medical, physical and psychological anthropometry this is not an idle dream—and let us allow none to enter who fails to reach this standard'. Jewish refugees who failed this test, as nearly all of them clearly would, might be encouraged to settle, instead, in thinly populated parts of the globe, where the local people were *below* their physical and mental level—perhaps Palestine. But Britain had a duty to preserve and improve its stock, and the unrestricted admission of

Jewish or indeed of other types of immigrants was hardly conducive to this end. Pearson and Moul hastened to add (as did the Whethams) that they were quite unconcerned with the *religion* of the Jews and had no anti-semitic prejudices.[34] But that, of course, had also been the claim of even rabid restrictionists, like Arnold White, during the Aliens Bill controversy.

Yet anti-semitism is not a charge that can reasonably be made against any but a handful of eugenists. Many Jews were prominent in the movement, like Dr Sidney Herbert, and contributions by Jews, written from a Jewish standpoint, were published in the *Eugenics Review*. There is no evidence whatever that Jews were ever made to feel unwelcome in the E.E.S. In March 1913 the Society set up a Jewish Committee to 'enquire into various questions directly connected with the Jews', a committee on to which Jewish members were to be co-opted.[35] But this should not be construed as evidence that eugenists had any hostile designs on the Jewish people. As even Pearson admitted, the Jews had much to offer other races in the matters of sex hygiene and race culture.[36] The eugenists especially admired the pride that Jews took in their family and their ancestry; as the Whethams put it, Jews seem 'to have had a very strong racial instinct, a profound sense of the importance of heredity', and the Mosaic code itself 'enshrine[d] many profound biological truths'.[37] In many respects the Jews were a model of what eugenists were seeking to establish: a closely knit community, which had identified religion with a sense of racial destiny and which invested its customary sexual and hygienic regulations with all the weight of religous authority. Just as eugenics seemed to some Jews to provide proof that their traditional beliefs and rituals had 'scientific' justification, so non-Jews could point to the Jewish community with satisfaction to demonstrate that eugenics was a practical proposition, a creed that could be absorbed into the social and emotional life of a whole people. To quote Arnold White, of all people: 'the existence of the Jewish race is a standing advertisement of the truth of the science that Francis Galton has revived . . .'[38]

Finally, the alleged 'purity' of the Jews made them a source of legitimate scientific interest to all concerned with eugenic problems. Investigations indicated that the incidence of disease was very different among Jews from what is was among non-Jews: tuberculosis, for example, being very rare, but defective eye-sight commonplace, facts that were partly explicable in terms of acquired immunity and partly to the results of in-breeding. The interest taken in the racial peculiarities of Jews was not always of a very intelligent kind; Dr Radcliffe Salaman's attempts to show that the familiar Jewish facial expression was a recessive character which 'Mendelised' are a case in point. But Salaman's conclusion, that it was 'essential that a pure stock possessing such qualities [as the Jews had] should be kept in existence' was one with which many, perhaps most, non-Jewish eugenists were heartily in sympathy.[39]

It remains to be asked why, holding these beliefs, they did not as a group postively encourage, rather than deprecate, Jewish immigration. Similar questions were raised at the time by liberal critics of immigration restriction. The answer can only be that the Jews entering Britain from Eastern Europe in the post-1880 period aroused the class hostility of many eugenists. While approving of Jews as a race, these particular Jews were destined to swell the ranks of the urban poor, perhaps to fall dependent on public funds. Even if they adapted themselves successfully to their new environment, they were likely to meet with disapproval, so long as they formed a closed community. Many eugenists seem to have believed that there could only be an improvement of the stock if pride in one's race could be kept vividly alive. An article in the *Eugenics Review* in January 1910 commended the Australians for being conscious of the 'menace of the yellow races'; this could only be 'a healthy influence', since 'the proximity of powerful and threatening neighbours has more than once in the world's history produced a nation of more virile and even heroic men'.[40] Commitment to the goal of a multi-racial society, it was presumed, would have the very opposite result.

The reference by the *Eugenics Review* to 'the menace of the yellow races' raises the further issue of what view eugenists took of non-European peoples. This issue can be quickly settled. With very few exceptions, eugenists simply assumed, in a very innocent and unselfconscious way, that the white races were biologically superior to the coloured races, although distinctions were made between 'advanced' Asiatic peoples like the Japanese, with considerable economic and military achievements to their credit, and extremely 'backward' races like the Negroes. This prejudice was, of course, very widespread in Edwardian England, and can be found in many otherwise 'progressive' liberal and socialist circles.[41] Perhaps for this reason British Eugenists seem not to have thought it necessary in these years to provide supporting evidence for what they regarded as a self-evident truth. But William McDougall, writing on the subject of 'psychology in the Service of Eugenics' in July 1914, anticipated a later phase in the movement when he advocated the application of intelligence tests to 'the various sub-races of mankind', in the hope that this would validate his 'intuitive' belief in racial inequality.[42] McDougall's hopes were very shortly to be fulfilled![43]

Meanwhile, the main problem to which British eugenists felt obliged to address themselves was whether marriages or sexual liaisons between members of widely contrasting racial groups should or should not be permitted. In 1911, a South African wrote to the E.E.S. for information on this point.[44] What reply, if any, the South African gentleman was sent remains obscure, but the general view among British eugenists was that miscegenation was a 'racial error', as the collapse of the Spanish and the Portuguese Empires eloquently testified.[45] Unions between closely related strains might strengthen the stock, but those between members of clearly differentiated racial groups, such as Europeans and Negroes, were likely to produce children lacking in physical, mental, or temperamental balance.[46] The leaders of the E.E.S. carefully avoided the wild racialist rant that Arnold White wrote for *The Referee.*[47] But Pearson was characteristically forthright. When asked by a journalist on the *Observer* for his views about 'Black and White marriages', he simply 'expressed his conviction that such unions "could only raise the blacks by lowering the whites"'.[48] Pearson also proclaimed that the Negro was one of those races which belonged 'to the childhood of man's evolution', a fact that justified Europeans, 'even for their own [the Negroes'] benefit, when we have suspended the stringent action of natural selection, in treating them as children'.[49] A more sympathetic approach to 'barbaric' peoples was consistent with attachment to eugenic principles. A. E. Crawley gave a sensitive description of sexual taboos among certain so-called 'primitive races', and concluded: 'As compared with the sentimental variety prevalent in civilised life, this early form of respect deserves the style of scientific. It is the respect of the medical eugenist rather than of the ethical or religious sentimentalist'.[50] There were passages in Galton's own work which might have stimulated further enquiries along these lines. They were not taken up. Most eugenists remained obstinately ethnocentric.

Yet this account of the 'racialist' strand in the British eugenics movement can be pressed too far. There were elements in eugenical thinking which prevented the elaboration of full-blooded theories of race. For a start, even a rudimentary understanding of genetics would have been enough to disabuse eugenists of the popular belief in the existence of 'pure' races.[51] At times they may have spoken and written as though they accepted the fixity of racial characteristics, but, as J. B. S. Haldane observed on a later occasion, this position was incompatible with the eugenists' firmly held conviction that racial deterioration had set in.[52] If pressed to choose between these two theories, few eugenists would have hesitated to discard the notion that races had immutable personalities. Moreover, eugenists were fond of stock-breeding analogies, and they knew that in the animal and plant world many successful new breeds had been created through 'crossing'. Whether two particular races should be 'crossed' would entirely depend, they believed, upon the peculiarities of the two races in question, but at least inter-racial marriages could not be dismissed out-of-hand as a biological error.

But more important than such considerations is the evidence that in pre-war Britain the 'racial issue' was not given much prominence; most of the wilder statements about, for example, Negro inferiority, which one encounters in British eugenical literature, turn out to have been made by Americans or to be based upon American 'research'.[53] In the inter-war years crude racialist gibes seem to have become slightly more common. But before 1914, at least, no serious attempt was made by any eugenist, apart from Pearson, to justify British rule in Africa and Asia on biological grounds. This may seem surprising, until one understands that British eugenists were not very much interested in what was happening overseas. In the 1908–14 period attention was, instead, focused on the alleged iniquities of the reforming Liberal Administrations of Campbell-Bannerman and Asquith.

5

Attitudes to Class and Social Welfare

We have seen that eugenists feared that the population was deteriorating because of the differential birth-rate. In earlier generations the higher mortality among the poor would have reduced the gap in effective fertility to very modest proportions. But, especially since 1891, the general death-rate had been falling steeply, presumably because of improved medical care, and this, eugenists gloomily observed, was permitting large numbers of the unfit to survive and thus transmit their defects to their offspring. So bitterly did many eugenists bewail the suspension of natural selection that they seemed to be advocating a return to unrestricted competition. This laid them open to the crushing rejoinder that the premises of eugenics, if true, would lead one to search for the Superman in the slums of the big cities where competition worked in all its primitive fury among masses of unskilled labourers engaged in a desperate conflict for a bare subsistence wage.[1] Moreover, as Sidney Webb observed, in human societies the question of who survived was 'determined by the conditions of the struggle, the rules of the ring'; 'where the rules of the ring favour a low type, the low type will survive, and vice-versa'.[2] However misleading their rhetoric, intelligent eugenists were well aware of these considerations. They entirely took Webb's point, but whereas the latter drew attention to the dangers of slum life, they themselves were more worried by the existence of generous State aid and numerous ill-organized charities, which together (so they claimed) had created an environment in which a 'reversed selection' was taking place—in which diseased, parasitic, and incompetent persons of various kinds were assured of a comfortable existence at the expense of the 'efficient', who were being taxed to support them and their numerous offspring.[3] This, in the eugenists' view, was the cause of physical deterioration and the multiplication of the unfit.

Unlike Spencerian Social Darwinists, eugenists did not favour a policy of minimum State intervention. They rather wished to replace natural selection by 'rational selection'. The teaching of biology was that man was a member of the animal kingdom; he had, however, to be converted from a 'wild' into a 'domesticated' animal, and this would involve an extension of State responsibility into hitherto unregulated areas of social life, like the procreation and rearing of children. The environmentalist case was not completely ignored. Eugenists were simply working for different political and social changes from those favoured by socialists and radicals: changes that would enhance the well being and, hopefully, increase the birth-rate of the 'efficient' middle classes, while reducing the numbers of the 'socially dependent'.

But eugenists thought that environmental influences were only important in so far as they encouraged or discouraged the spread of particular human and social attributes and determined which of these would be 'selected'. They rejected the belief that environmental agencies could *cause* the human species to be modified in either a beneficial or a detrimental sense. That acquired characteristics were inherited was a theory which few biologists still held. The researches of August Weismann, and his views about the immutability of the germ-plasm, implied that while use, or accidents, or nourishment, or other kinds of stimuli might modify the somatic cells, they would not affect the germ-cells, so that whatever advantages an individual might have reaped in his own life-time would die with him; they could not be transmitted to his descendants. More than one practical deduction could be drawn from this scientific proposition.[4] But eugenists obviously used it to throw cold water on educational effort and social reform generally. Not only could the species not be improved by these agencies, worse still, these agencies might have positively harmful consequences, since they would tend to 'mask' evidence of biological degeneration. Indeed, eugenists used this argument to explain why insufficient notice was being taken of national deterioration.

Thus, philanthropists, social workers, and educationalists had been busily fussing over the *symptoms* of a much more deep-seated problem, which derived from factors of genetic endowment. Eugenists mocked at Medical Officers of Health and schoolteachers who had started from the premise that, given a constant environment, all human beings could perform at an equally adequate level: this fallacious view had produced nothing but misery and heart-ache. Dedicated teachers had tragically wasted the best years of their lives on children who were congenitally incapable of responding to their efforts. Pearson produced a parable about a man trying to make a razor; although a good razor required tempering and setting before it could function efficiently, its cutting edge would only be sharp if the material were sound steel. Similarly with education. One could not produce brains by multiplying universities and technical colleges, although such institutions were useful to men and women of high in-born intelligence.[5] Doctrinaire refusal to attend to the requirements of heredity had involved real hardship for dull children. Displaying for once a deep humanitarian concern, eugenists lamented the lot of a working-class child forced to undergo an educational process to which he was utterly unsuited.

Leonard Darwin was anxious that eugenists should not make enemies needlessly, and sometimes tried to discount the impression that eugenics was an *alternative* to social reform. The question of whether heredity or environment did more to mould the individual he likened to a debate between farmers as to which was the more important, manuring or ploughing; obviously *both* were important, and if eugenists concentrated upon the former this was because no other pressure-group existed which could agitate on behalf of the unborn and hold a watching brief for the well-being of the race.[6] But in general, the whole tenor of eugenical literature was to demonstrate within what narrow limits the social reformers were operating. The work of Pearson and his colleagues at the Biometrics Laboratory was designed to measure the correlation between certain human attributes on the one hand, and environmental and hereditary factors on the other. The evidence satisfied these workers, if not everyone else, that in general the intensity of inheritance was some five to ten times greater than that of environment.[7] Pearson could thus ascribe the social problems of contemporary Britain to one simple factor: 'we have placed our money on Environment, when Heredity wins in a canter'.[8] In an often cited passage, the moral was drawn: 'the first thing is good stock, and the second thing is good stock, and the third thing is good stock, and when you have paid attention to these three things, fit environment will keep your material in good condition'.[9]

Environment only seemed to be so important because of the slipshod methods of most social scientists. Medical Officers of Health, like Arthur Newsholme, produced statistics which apparently proved that child death-rates were higher among the residents of back-to-back houses than among those inhabiting a more salubrious neighbourhood. But correlation was not the same thing as causation. Pearson mocked at Newsholme's use of statistics by using an analogous method to 'prove' that the rise in the cancer figures had been 'caused' by an increase in the importation of foreign apples. All this ignored, so the eugenists contended, the important point that men made their environment, and were not simply moulded by it. In the words of Edgar Schuster: the eugenist had to 'find out what qualities lead different individuals into different environments. We must be careful not to assume that the environment is thrust haphazard on us, for it is largely moulded by our own characters'.[10] At its extreme, this argument could be extended to show that the slums were themselves biological phenomena; they were created by a distinctive biological type, the 'chronic slum dweller', and only when this undesirable type had been bred out of the race would it be possible to effect a lasting improvement in housing conditions. If, as Heron thought possible, 'the mentally and physically inferior parents gravitate to the inferior environment',[11] then most social workers were clearly working along wrong lines.

Such conclusions would be unpalatable to many people, but eugenists warned of the dangers of being carried away by an unreflecting humanitarianism. Pity was an instinct that needed to be guided by reason and biological science. In Pearson's words, 'one factor—absolutely needful for race survival—sympathy, has been developed in such an exaggerated form that we are in danger, by suspending selection, of lessening the effect of those other factors which automatically purge the state of the degenerates in body and mind'. He added, 'we cannot go backwards a single step in the evolution of human feeling!

But I demand that all sympathy and charity should be organized and guided into paths where they will promote racial efficiency, and not lead us straight towards national shipwreck'.[12] From this perspective much philanthropy could be dismissed as a cowardly escape from harsh reality; true courage required the disciplining of one's mere feelings and a deliberate subordination of short-term emotional satisfaction in the interests of the race as a whole.

The crippling financial burdens being imposed upon the 'efficient' members of society by thoughtless social reformers was another theme upon which eugenical writers were constantly harping. Major Darwin tried on several occasions to 'cost' the efforts being made by the community to keep the 'unfit' alive and comfortable. He concluded that the amount of money spent on law, justice, the police, the relief of the poor, infirmaries and lunatic asylums, 'all services which would be much less needed if the unfit were eliminated', came to approximately £48 million a year. This left out of account the cost of running special schools, expenditure on National Insurance, and the vast sums of money, perhaps £10 million a year, subscribed to charitable bodies.[13] But the 'costs of degeneracy' did not end there. The one million paupers in daily receipt of relief were being looked after by 'a whole army of able-bodied officials and attendants'; in a Eugenical State most of these skilled men would be 'set free from their economically useless occupations' and 'employed instead on productive services'. Moreover, the elimination of the pauper and the wastrel would immensely increase the productivity of the work-force, as had been shown by the researches of Munsterburg and others on Industrial Efficiency.[14] The whole community would benefit from this, including honest workers at present penalized by the presence in the ranks of the work-force of shirkers and weaklings.[15]

But, most important of all from the eugenic standpoint, a reduction in the numbers of the unfit would automatically lead to a reduction in taxation. This was seen as important, since many eugenists held the implausible view that it was the punitively high level of taxation which was preventing middle class couples from having large families. With the standard rate of income tax a mere 1*s.* 2*d.*, even after the People's Budget, and the higher rate of surtax only 9*d.* in the pound, this argument bears little serious consideration. But it was customary, nevertheless, for eugenists to speak as though the country had already reached the limits of its taxable resources. Montague Crackanthorpe berated the Socialists for attempting to stir up class enmity; let them remember, he said, 'that the so-called "rich" are, in the majority of cases, saddled with such heavy obligations for the benefits of others that what is left over for themselves is but a very modest sum'.[16] He, too, thought this was why middle-class parents were practising family limitation. There was an obvious political appeal in this line of argument at a time when the Liberal Government was embarking upon an expensive social reform programme, which, modest though it may appear in retrospect, did represent some kind of threat to the wealthy and privileged. A letter to the *Sheffield Telegraph* in March 1910 illustrates this point clearly enough; after protesting at length about the 'murder' of the best stocks in order to provide a suitable environment for the worst stocks, the writer concludes: 'Therefore, I say again—hands off the rates and hands off the taxes. What is needed is a complete reversal of the rotten Socialistic policy which has done, and is doing, so much mischief in the country . . .'[17]

Frightened middle-class groups would also have been attracted to eugenics for the reason that it purported to give a scientific explanation, and hence a moral justification, of class divisions. Class stratification was portrayed by many eugenists as a necessary consequence of biological evolution, which had many parallels in the animal and plant world. No less an authority than the geneticist, William Bateson, was on record as saying: 'As the biologist knows, differentiation is indispensable to progress. If the population were homogeneous civilisation would stop'.[18] The vulgarity and ignorance of nineteenth century educationalists and social reformers was nowhere more clearly seen than in their tirades against the arbitrariness and injustice of class distinctions. In fact, eugenists argued, classes were biological phenomena. Through isolation and inter-marriage, distinct human sub-groups had come into existence, with marked physical features and mental traits, which were inborn not acquired. When visiting a cattle market Bateson perceived as many types among the human beings as among the animals they were managing, the only difference being that men were the offspring of almost random mating, and so had not formed separate breeds.[19] The Whethams went further, when they wrote: 'the variations of type among us, as indicated by the different social strata, show the existence of variations of innate physical and mental characteristics as real though infinitely more elusive than the differences between, for example, the Highland cattle and the Guernsey cow . . .[20] Not only social classes in the broad sense, but even occupational sub-groups were often treated in this way. To quote Pearson: 'The differentiation of men in physique and mentality has led to the slow but still imperfect development of occupational castes within all civilised communities'. He added that 'in a perfectly efficient society, there would always be castes suited to specialised careers—the engineer, the ploughman, the mathematician, the navvy, the statesman, the actor and the craftsman', and people would largely marry others of the same caste.[21]

But this assumption that the class system expressed differences of genetic endowment was precisely what infuriated the critics of eugenics. What evidence, they complained, was there for supposing that wealth or social status were closely correlated with eugenic worth? In reply, eugenists usually admitted that it would be wrong to identify genetic factors with social status, and denied that this was what they were doing. However, the qualification once made, they invariably forgot all about it, and blithely proceeded to make the very claim that their critics had rightly challenged. The fact remains that all but a small minority of 'reform eugenists' held the unshakeable conviction that, broadly speaking, the upper classes were superior to the lower orders in all those attributes to which humans attached value: health, sturdy physique, and intelligence.

Opponents were not convinced. Even if the existence of these differences of physique and intelligence could be demonstrated, they said, this would only indicate that the socially privileged enjoyed advantages which working men lacked. If there were genuine equality of opportunity, the working classes, or many members of this class, would emerge as at least the equals of their social betters; so let there be reforms in education and the 'abolition' of poverty, before hasty judgments were made about the inborn qualities of particular social classes. This line of argument sometimes got home. Even a rigid hereditarian like E. J. Lidbetter could write: 'We must take steps to ensure that in no phase of our social life can it be truly said that the environment of the people is so bad' that a question mark rested over the importance of genetic endowment.[22]

But the dominant theme was the one which Francis Galton himself had bequeathed to his followers. In *Hereditary Genius*, Galton argued that men who achieved eminence and those who were naturally capable were to a large extent identical.[23] The highly gifted possessed a kind of nervous energy and capacity for sustained effort which carried them through whatever obstacles society might place in their way. And because he believed that the upper classes of Britain were being 'largely and continually recruited by selections from below', they had by far the lion's share of natural ability, while 'the lower classes [were], in truth, the "residuum"'.[24]

The provision of free elementary education, scholarship ladders, and the like seemed to have strengthened Galton's case. Edwardian eugenists believed, however mistakenly, that the social structure was now so fluid that any bright lad could rise out of the working class and make his way in the world. There was a continual 'sifting out' of ability, with the able rising into the upper classes, to replace those whose lack of natural capacity had led to their social demotion. This process, they said, could not be taken much further. In McDougall's opinion, 'we have now well-nigh perfected the social ladder'![25] It was in order to confirm this hunch that McDougall, then at Oxford University, encouraged the young Cyril Burt to devise tests which could measure 'general intelligence'. These tests were first carried out on two groups of Oxford schoolchildren, one attending a 'high-class Preparatory School', the other a 'superior Elementary School'; they showed that those of higher social status did consistently better; thus Burt was able to conclude that 'the superior proficiency at Intelligence Tests, on the part of boys of superior parentage, was inborn'.[26] British eugenists were slower than their American counterparts to grasp the relevance of intelligence testing to the Nature/Nurture controversy, but by 1914 they were beginning to play an important part in eugenical propaganda.[27]

But even without the aid of intelligence tests, most eugenists were satisfied that by and large people were in the social class to which their inborn qualities entitled them. This is evident from the frequently made assertion that in general, wages provided an accurate measurement of individual efficiency. Leonard Darwin thought that both intelligence and scholastic tests were impractical for the general purposes of eugenical reform. But, he argued:

> The qualities which make a man able to support his family are all of racial value; and the rate of wages . . . may be made to afford some indication of the value of the innate qualities of the wage-earner. The ill paid are, no doubt, very often superior to the well paid; but the correlation between wages and good qualities is likely to increase continually as time goes on. That such a correlation does exist is rendered highly probable rather than actually proved by several comparisons between social strata which have been made by means of intelligence tests . . .[28]

But on Darwin's own admission the coincidence between eugenic worth and income (hence social status) was by no means complete. Let the remaining 'anomalies' be swept away as soon as possible, he argued. But he added, let it not be supposed that advances made in the direction of equality of opportunity would lead in the long run to *less* inequality; the opposite was more likely to occur, because reform would simply reveal more starkly the vast differences in natural endowment with which human beings came into the world.[29]

To most eugenists the prospect of an unequal and hierarchically organized society held no alarm, provided that the differentiation was made on a eugenic basis. William Bateson believed that the feudal system approximated as closely as any man-made contrivance had yet done to the natural order. Society, he added, was returning to a similar sort of structure, although this time it would have a scientific foundation. Like most eugenists, Bateson was worried by the social turbulence and industrial agitation that marked the pre-war years. But he thought that his new feudal society would greatly reduce the tension, because in such a society the different grades of human being would all occupy their rightful place. 'At such a time as the present much of the intensity of discontent is due to the fact that some are at the bottom who should be higher, while some are high who should be lower'.[30]

Thus, like Darwin, Bateson wanted there to be a revision of social status. But few eugenists seem to have supposed that this would entail a radical shake-up of the entire social system of their day. Indeed, many of them were opposed to even modest reforms that might increase social mobility. Perhaps we should summarily dismiss, for the nonsense that it was, Mudge's extraordinary assertion that the scholarship boy from a poor background carried, as it were, latent working-class genes and so would tend to produce offspring who would 'revert'.[31] More rational were the fears of the Oxford philosopher, Schiller, that precisely because the working classes were being 'drained' of their ablest stocks by scholarship ladders and the like, there would be few if any men and women of transcendent ability who could come to the fore should there by a national catastrophe or a social upheaval similar to the industrial revolution. The more thoroughly ability was segregated into a ruling class, the greater the dangers to which the nation was exposed.[32] A variation on this lament was Darwin's fear that the 'residuum', baffled at their own incapacity, might by sheer force of numbers eventually overwhelm the efficient, and that unless they were led by able people from their own ranks, who had at least some inkling of the value of science and the arts, the lower orders might one day lash out at the very fabric of civilization.[33]

In addition, there was the argument of the Whethams that the bright working-class child who went up in the world tended to marry either unsuitably, or too late, or perhaps not at all; in eugenic terms this was a grave evil, because it meant that stock of sound worth was being permitted to die out. 'Better that an able carpenter should develop slowly into a small builder, leaving six tall sons to play their part manfully, and, perchance, rise one step more, than that he should be converted by a County Council scholarship into a primary schoolmaster, or second-grade Civil Service clerk, and that there the usefulness to the race of the innate abilities of which he is the temporary trustee should cease for ever'.[34] One would have thought that the difficulty being complained of was a social one, capable of being met by some appropriate social or political arrangement, but as the above extract clearly shows, the Whethams, like other eugenists, were strongly predisposed, for reasons that had nothing to do with biology, in favour of preserving the *status quo*, and so disliked an educational system which enabled some working class boys to escape from their due station in life. Other eugenists shared the Whethams' fears that modern education was to blame for much of the political unrest of the day. Significantly, Darwin wanted all those who could benefit from a good education to be given the chance, but he added that education should usually fit children for the way of life pursued by adults of the same class, and that parents should be told that 'the possibility of their children rising to a social station higher than their own' was 'a not improbable event, but one entirely out of the parents' control'.[35] As J. F. Tocher succinctly put it: 'our object as reformers is to work towards a state of Society where the action of individual units, as a whole, shall produce stability in the community'.[36]

So far eugenics has been presented as a creed with highly conservative implications, as indeed it was. But there were features peculiar to eugenics which marked it off from other conservative ideologies. This will become clear if we examine the way in which eugenists treated the landed aristocracy and the peerage. A few eugenists went out of their way to support the traditional 'upper classes'. The Whethams' writings were all designed to justify, on biological grounds, 'our best families', and to help save them from the threat of 'extinction'. 'The old governing classes of England, as of other similar nations', they wrote, 'incorporate an instinctive sense of public duty and acquire a large share of the natural aptitude for administration'.[37] It will be noted that the Whethams were perfectly prepared to invoke acquired aptitudes and family traditions when it suited their purposes, although they contemptuously dismissed such considerations as 'fallacious Lamarckianism' when employed by radical social reformers. Such was the anxiety of the Whethams to justify the special privileges and influence of the landed aristocracy, and its links with the church, the armed services, and the offices of State. 'In no department of knowledge', they wrote, 'was the self-satisfied individualism of 1850 more positive or more at fault than in matters of genealogy and heredity. The aristocratic theory of the family, which even those who believed in and acted on hardly then ventured to support openly, contains the root of the matter, and only wants restating in modern terms to take its place as a great scientific truth'.[38] These sentiments were expressed in the Whethams' book, *The Family and the Nation*, published in 1909, the year of the People's Budget—the political significance of this hardly needs underlining.

The Whethams were not alone in employing eugenical arguments in support of the British landed classes and the peerage against their radical assailants. Rather unwisely, perhaps, Montague Crackanthorpe permitted himself some quite open political sallies in his 1910 Presidential Address. Along with other indiscretions, he praised the House of Lords as an assembly largely composed of men who had achieved distinction in one walk of life or another and of the descendants of such men, whom one would also expect to be well above average in ability: 'This is probably what Lord Cromer meant when he said the other day, speaking in his place in Parliament, "It is very easy to go too far in condemning the hereditary principle; there is something in Heredity".'[39] But Galton and Pearson were most irritated at remarks like this which confused theories of heredity with the principle of primogeniture. Moreover, they had other reasons for doubting the worth of a second chamber such as Britain possessed. For a start, as Pearson pointed out, it was often forgotten that a man had 'sixteen grandparents, and, possibly, only one of them may be of distinction, the man who won the title'. The mathematical probability of this exceptional ability being passed down for several generations through the eldest son would obviously be quite small.[40] Moreover, Pearson's statistical researches led him to the controversial conclusion that the first-born child ran a greater risk of coming into the world with such defects as tuberculosis and insanity than subsequent children. If true, this suggested that the deliberate restriction of births, and especially the growing prevalence of the one-child family, portended national disaster.[41] But it also told very heavily, of course, against a hereditary chamber recruited by primogeniture. Galton, who agreed with Pearson on this issue, wrote a letter to *The Times* in March 1910, pointing out that 'the claims of heredity would be best satisfied if all the sons of peers were equally eligible to the peerage, and a selection made among them, late researches having shown that the eldest-born are, as a rule, inferior in natural gifts to the younger-born in a small but significant degree'.[42] Pearson himself, one presumes, took a very real pleasure in 'debunking' the House of Lords and ridiculing its pretensions; he still retained a few attitudes from his socialist past which brought him satisfaction in iconoclastic attacks on the established social and political order.

The same cannot be said for the majority of eugenists. Yet, leaving aside the Whethams, most of them obviously viewed the landed aristocracy with mixed feelings. Even the Whethams were forced to admit that, once a man had achieved the social recognition conferred by a peerage, the pressures of selection relaxed, and there was a danger of the children of the successful man slipping backwards: 'When a family becomes firmly established among the upper classes, the pressure of selection becomes less acute. Places are found for the sons, whether their abilities deserve them or not; some of the daughters make good marriages, regardless of whether they possess their share of the family ability. Selection ceases to a great extent, and reversion to a lower level inevitably occurs'.[43] But for this, they suggested, the human species would have become much more differentiated than it actually was. Obviously this was, from the conservative viewpoint, a damaging admission.

Other eugenists went further. In eugenical literature there are often expressions of disapproval of inherited wealth; wages may often have been presented as an indicator of individual efficiency and merit, but only very rarely are similar claims made for wealth.[44] A precursor of the eugenics movement, Dr Haycraft, produced the slogan, 'Property Holders Less Capable than Property Acquirers'.[45] Socialists, indeed, were able to argue that property itself must be a dysgenic institution in so far as it interfered with natural selection. Without going that far, eugenists tended to look with some suspicion on the landed aristocracy. This class undoubtedly contained families where a strong sense of public service still ran strong; these men were the true aristocrats, as well as being so in title. But not even Hilaire Belloc had more scathing things to say about the way in which the existing ruling class was turning itself into that most despicable of social groups, a plutocracy. Luxury, so ran the argument, was physically and morally debilitating. What was worse, from the eugenic point of view, it led able men and women to make inappropriate marriages, because wealth was valued more highly than sound stock. Where there was a spirit of self-indulgence, people would neglect their racial responsibilities by limiting their families, the women through an ignoble shrinking from physical pain, the husbands because the expense of bringing up children might interfere with their creature comforts. This was a temptation constantly before an aristocracy, most of whose members had abandoned their old traditions of ascetic patriotism. In Schiller's caustic words: 'Satire has often noted that the sole merit of the *grand seigneurs* [*sic*] was merely *de se donner la peine de naître*; and nowadays even this appears to be becoming too much trouble for them or for their parents . . . it is precisely because these favourites of fortune already have what most desire, *and have to work for*, that they degenerate . . . our present sham nobility . . . has become a social institution that means nothing biologically'.[46]

In addition, eugenists had learned from Galton that peers who attempted to rehabilitate their family fortunes by marrying rich heiresses ran a significant risk of procuring a barren wife. Only the Whethams chose to suggest that perhaps the time had come for returning to the old practice of combining a hereditary title with a grant of lands or a substantial sum of money so as to do away with the temptation of peers marrying heiresses for their worldly goods.[47]

But there were less expensive ways of 'purifying' people's attitude towards marriage. Bernard Shaw thought that the best way of improving the human race was to institute equality of incomes, so that, undeterred by arbitrary class distinctions, men and women could choose a marriage partner, without consideration of rank or wealth, in accordance with the promptings of instinct, alias love, that very effective eugenic agency.[48] Put less flamboyantly, this was the 'biological' justification for socialism which had long been urged by Alfred Russel Wallace, the co-discoverer of natural selection.[49] Saleeby announced that, though personally no socialist, this argument struck him as being 'incomparably the best argument for that creed; and if it were proved that only through socialism could the utmost be made of women's choice of husbands, then no argument against socialism could have any appreciable weight at all'.[50] But Shaw and Wallace, as socialists, believed that the existing class system was a mere lottery, and not even Saleeby could accept a critique of society that went to these lengths. Eugenists usually contented themselves with advising their readers to abandon the ignoble search for wealth at all costs and set an example of austere living and family responsibility that others could follow. As Darwin put it: 'Example being better than precept, those parents who put aside both social ambition and all useless displays of wealth, and who follow the sound rule of never making friends with persons they cannot respect, will be making households where their children when young will naturally absorb high ideals and when grown up will be likely to meet with youth of good stock'.[51] Eugenists were fond of simple-minded moralizing of this kind.

But luxury and flamboyant displays of wealth were obviously not the prerogative of the landed aristocracy. Eugenists also had harsh observations to make on this score concerning the entrepreneurial and business classes. Speculators were particularly frowned upon, while Bateson told the British Association in 1914 that although capital was an 'eugenic institution', 'the rewards of commerce are grossly out of proportion to those attainable by intellect or industry. Even regarded as compensation for a dull life, they far exceed the value of the services rendered to the community . . .'[52]

Who, then, were the heroes of the play? In a word, they were the professional middle classes and the intelligentsia, or at least that section of it not corrupted by an effete humanitarianism. Threatened from above by the frivolity and pretentiousness of 'high society' and from below by an ignorant unwashed populace, the intellectual middle classes saw themselves as those who did the really important work of the world, for inadequate recognition and reward. Typical of eugenists was Schiller, who denigrated both upper and working classes in order that the professional man might stand revealed for what he was, the most perfect specimen yet produced by human evolution.[53] This bias had been present in eugenical literature from the very start. A famous passage in Galton's *Hereditary Genius* runs: 'The best form of civilization in respect to the improvement of the race, would be one in which society was not costly; where incomes were chiefly derived from professional sources, and not much through inheritance; where every lad had a chance of showing his abilities, and, if highly gifted, was enabled to achieve a first-class education and entrance into professional life, by the liberal help of the exhibitions and scholarships which he had gained in his early youth . . .'[54]

That the eugenics movement was pre-eminently concerned with the welfare of the professional classes is shown not only by the literature it produced, but also by the membership it attracted. Of the 634 members of the E.E.S. in 1914, a high proportion consisted of scientists, medical men, university lecturers, and men of letters; these, at least, provided the active membership and gave the Society its particular tone. It is noticeable, too, that nearly all the affiliated branches were located in university cities, and clearly depended very heavily on the participation of university faculty.

Between eugenists and the working-class population, on the other hand, a great gulf existed, and nothing in eugenical literature is so striking as the way in which working people are frequently discussed as though they were denizens of some other planet.[55] But eugenists have been accused not only of being ignorant of and insensitive to the working-class community, but also of elaborating doctrines which 'amounted to a practice of culling the socially and economically deprived'.[56] To this charge eugenists would have replied that they had no hostility to working-class people as such, and would also have readily conceded that the 'fit' members of this class possessed racial qualities of the highest value.

But they insisted upon the distinction between 'fit' and 'unfit', and often spoke as though this involved differences of kind rather than degree. E. J. Lidbetter argued that, next to the segregation of the feeble-minded, the most important eugenic problem was to prevent the crossing of the line between the 'fit' and 'unfit' sections of the working class; the prevention of this, he said, would 'check, and ultimately bring an end to, that exchange between strength and defect which at once perpetuates the defective stocks, and vitiates the good stocks, by the marriage inter-change which is constantly going on'.[57] During the period of the Great Labour Unrest immediately preceding and following the First World War, some eugenists conceived the idea of breaking the alarming cohesion and militancy of trade unionists by urging the 'artisans' not to be so foolish as to fight for privileges for the 'inefficient' members within their ranks; could not the working man of good physique and intelligence see that he would be the prime sufferer from a socialistic system in which loafers, wastrels and 'unemployables' were kept in comfort by support from public funds?[58]

The 'unfit' were thus sometimes identified with the mass of unskilled workers, but usually more narrowly with the socially dependent, men whose distinguishing feature was their inability to maintain an independent existence. Even the latter group, eugenists admitted, contained many who were merely 'unlucky'; the community had a clear duty to help, in no patronizing spirit, such unfortunates as those thrown out of work by cyclical unemployment, or women and children thrown on to the poor law by the sudden death of the breadwinner.[59] Having made this admission, eugenists went on to claim that in fact the bulk of the socially dependent were deficient in some way or another. These were the barbarians within the gates.

It must be remembered that the E.E.S. had been formed in late 1907, at a time when unemployment was moving sharply upwards, reaching a peak in September and October of the following year. The Labour Party's repeated introduction into the Commons of a 'Right to Work' Bill had drawn further attention to this grave social problem. After 1910 the level of unemployment dropped, until in April 1913 it comprised only 1·7 per cent of all trade unionists. But although unemployment largely disappeared from the headlines, the related problems of pauperism and destitution did not, thanks to the efforts of the Royal Commission on the Poor Laws, whose Reports came out in early 1909, and to the Webbs who kept up a lively campaign aimed at the abolition of the whole Poor Law system. It was into this debate that the eugenists eagerly plunged.

Eugenists viewed with contempt what they thought was the evasion of the real issues involved. William Dampier Whetham, addressing students at Trinity College, Cambridge, in January 1910, observed: 'Complicated questions of economics and sociology are only confused when they are discussed in the political arena. Please clear your minds as far as possible of all you may have heard, read or said on these subjects during the past few months, and look with me at the facts'. Whetham's contention was that neither the Majority nor the Minority Report merited much respect, since neither gave consideration to the evidence that pauperism and unemployment were in large measure due to inherent defect.[60] Incidentally, both Galton and Darwin had hoped that Pearson might be asked to offer 'hereditary' evidence before the Poor Law Commission, but Pearson waited in vain for a summons.[61]

The Eugenics Education Society subsequently attempted to fill the gap, by carrying out its own enquiry into the problem of destitution, with the aid of relieving officers from three workhouses.[62] The leading spirit in this investigation, Lidbetter, was himself a poor-law officer in the East End of London, who devoted much of his life to the laborious compilation of pauper pedigrees. In the most painstaking manner, Lidbetter built up evidence which pointed to the not surprising conclusion that individuals in receipt of poor-law relief usually had relatives who were similarly circumstanced. His pedigrees were actually quite compatible with an environmentalist theory of destitution, but Lidbetter himself was confirmed in his belief that there existed a hereditary class of 'chronic paupers' and unemployables, whose members were being steadily increased by the short-sighted activities of philanthropists and politicians. For how could 'bad environment' be to blame, when the environment had been continuously improved at great expense over the past fifty years, while the 'vast army' of paupers had shown no real sign of diminution?[63]

In Lidbetter's view, these pathetic creatures, drifting through life without purpose or any sense of social responsibility, were 'non-members' of society, a caste cut off from the world of independent men and women.[64] Incapable of responding to the appropriate stimuli, these 'hereditary paupers' not only burdened civilization, but also menaced its very survival through their high fertility.

But what was to be done with this 'social problem group', to use the phrase which the Wood Committee succeeded in popularizing in the 1920s? The first difficulty was in identifying them and separating them off from the merely 'unlucky' paupers. Were these 'unfit' members of society to be defined in administrative terms as persons in receipt of public assistance, or did they have actual defects of a kind that doctors could diagnose? Eugenists inclined to the second view, as can be seen from their habit of listing criminals and paupers with the insane, the tuberculous, and epileptics, when discussing the composition of the 'unfit'. Of course, Lidbetter was quite correct when he pointed out that a high percentage, perhaps eighty per cent of indoor paupers, were suffering from some kind of illness that incapacitated them for work;[65] in so far as those illnesses were of a hereditary kind, it could be argued that the reasons for these paupers' destitution were 'hereditary', although it was rather playing with the meaning of words to go on and claim that therefore they were 'hereditary paupers'.

Another large category within the ranks of the socially dependent were the feeble-minded; the Royal Commission on the Feeble-Minded estimated that they comprised about five per cent of the habitual inmates of casual wards, cheap lodging houses and night shelters.[66] Eugenists could quote the Commission itself in support of their demand for the segregation of the feeble-minded from the rest of the community. But they often went further, and implied that many 'unemployables' were border-line, feeble-minded people, of too low a mental type to earn an independent living, but not sufficiently retarded to be placed in custodial care. In Lidbetter's words, 'the danger arising from this condition of affairs is the more serious in that it is not recognised'.[67] Similar sentiments were expressed by American eugenists, who perhaps showed even greater enthusiasm than their British counterparts in tracing back delinquency, crime, and pauperism to the root cause of feeble-mindedness. As Haller remarks, the fact that these dependent people were 'so close to normal that [they] would pass for normal if not discovered by the Binet tests, made [them] all the more dangerous to the community'.[68]

But defective intelligence did not of itself explain what was wrong with the habitual pauper or unemployable. In an attempt to formulate what they meant, eugenists fell back upon the language of moral condemnation. There was something defective, they said, in the moral outlook of the 'unemployable'. Thus, Pearson spoke of the need to segregate those whose 'social inefficiency' stemmed not so much from mental inadequacy as from the fact that they took 'a view of life which [was] in distorted perspective' and were 'out of harmony with their economic or social surroundings'.[69] And the *Eugenics Review* created some mirth among its critics when it characterized the 'hereditary pauper' as a person who 'was born without manly independence'.[70] Edward Brabrook was even more severe: 'It is impossible to be entirely blind to the fact that large numbers of the chronically unemployed have brought their troubles upon themselves. There are the criminals and semi-criminal classes whose records keep them out of work. There are the drunkards, gamblers and incorrigible idlers. These are those not positively vicious, but who are incurable loafers'.[71] These ne'er-do-wells often gave themselves away by their physical appearance; 'a furtive glance and a low forehead', were often encountered among this class.[72] And yet, paradoxically, at the same time as showering the 'chronic pauper' with moralistic abuse, eugenists insisted that, coming as these people did from 'degenerate stocks', they could no more alter their behaviour than the leopard could change its spots. 'The charge of being improvident, often brought against this class, is a true charge, but it is so because of the people's incapacity to be otherwise', wrote none other than E. J. Lidbetter.[73]

The Webbs once observed of eugenics that it was 'in fact, just now the most fashionable kind of *laissez-faire*'.[74] Although true in one sense, this assertion conceals the sharp disagreements that divided eugenists from many orthodox defenders of *laissez-faire* doctrines. The Charity Organization Society [hereafter C.O.S.], for example, viewed the new-fangled creed with suspicion, because of the eugenists' tendency to undervalue the ethical principles of 'co-operation' and 'social service'. C. S. Loch wrote: 'We have to rely on the efficacy of home education and control and the many educational means at our disposal—schools, bands of hope, clubs etc., etc., and "good surroundings". We have also to rely on an administration of assistance guided by principles that lead to self-reliance and self-support'.[75] The C.O.S. were indignant at the suggestion that most paupers were congenitally incapable of responding to the stimulus of encouragement and incapable of improving themselves. Perhaps irked also by eugenists' contemptuous treatment of the Poor Law Majority Report, which some of their leading members had helped to draft, the C.O.S. defended the very poor from their new middle-class detractors, showing in the process a warmth of sympathy for which their Society was not exactly famous.[76]

Eugenists and the C.O.S. could agree, however, in attacking nearly all the social reforms introduced by the pre-war Liberal Governments. Old age pensions, for example, had few advocates in eugenical circles. The Whethams thought that the need for this form of State endowment had only arisen because of the declining birth-rate among respectable citizens. 'Now, as he grows old and past work, [the artisan's] maintenance becomes a well-nigh intolerable burden on the one or two children, who, possibly with families of their own to rear, suffer acutely, largely owing to the suppression of the brothers and sisters, who would have shared the responsibility with them at this juncture'. Yet, in the Whethams' view, the increased taxation necessitated by old age pensions would depress the birth-rate still further, so that a vicious circle was being created.[77] Karl Pearson employed a different argument: 'When we regard the present six or seven million pounds a year—soon to be ten or more millions—given to a mere environmental reform, which applied long after the reproductive age cannot possibly produce any *permanent racial* change, how deeply one must regret the want of knowledge and statesmanship, which overlooked the naturally disastrous policy of the factory acts, and did not seek its opportunity to endow parentage rather than senility with those annual millions'. Instead of this 'hasty vote-catching legislation', wrote Pearson, the government should have followed the German example, with benefits determined by eugenic considerations.[78]

Since the 1911 National Insurance Act equally failed to meet Pearson's criteria, it is not surprising that eugenists should also have expressed their doubts about the advisability of this measure. But though Bradbury prodded the E.E.S. into setting up a working party on the subject in 1912,[79] members seem to have had some difficulty in formulating an agreed position, and the *Eugenics Review* never carried a full feature on the National Insurance Act as it had done on the Poor Law problem. Saleeby was effusive in his praise for the Act in so far as it attempted 'to care for the last fortnight of expectant motherhood or ante-natal nurture'.[80] In general, the principle of a maternity benefit was welcomed,[81] although some eugenists felt it resembled too closely for comfort the Fabians' proposals for 'the endowment of motherhood'. But the sanitorium benefit was singled out for unfavourable mention by Mrs Gotto when addressing the International Congress of Health in Paris in 1912; the £1½ million made available for this purpose, she said, was a tax on the fit, which 'in the present dysgenic state of public opinion, [would] allow a larger number of [the tuberculous] to reach maturity and reproduce their kind'.[82] The £57,000 earmarked for research under the terms of the act met with approval, of course; in fact the Council of the E.E.S. had their eyes on the fund, some of which they would have liked to see spent on a study of heredity.[83] In general, however, eugenists did not like the National Insurance Act. It spread the risks of sickness and unemployment throughout the community in a way which they thought likely to stimulate the fertility of the 'unfit', while increasing the fiscal burdens which weighed upon the efficient; and it was a measure concerned only with mitigating individual suffering, which by-passed the causes of social distress and took no account of the long-term interests of the race.

6

Eugenics and Party Politics, 1908–14

It might seem, therefore, that eugenics provided an ideological prop to the more traditional members of the Conservative Party. However, many of its advocates, especially the scientists, viewed it rather as a message which transcended what ordinary people meant by politics; so overwhelmingly important was it, conventional political issues must pale into insignificance by comparison. Science had taken the place once occupied by an unscientific politics and ethics, and the latter would soon arouse interest only as historical curiosities. Saleeby spoke of eugenics as the touchstone by which the usefulness of all political panaceas could be tested: 'the question is not whether a given proposal is socialistic, individualistic or anything else, but whether it is eugenic . . . I claim for eugenics that it is the final and only judge of all proposals and principles, however labelled, new or old, orthodox or heterodox'.[1] Some of the scientists expressed amazement that intelligent people could still get excited over such trivialities as tariffs, dreadnoughts, religous education in schools and the other preoccupations of the party political machines; of what importance were battleships beside babies? And what purpose could be served by improving and perfecting the country's educational system if racial deterioration had once been allowed to set in? William Bateson, so his wife informs us, believed that Parliamentary politics were without foundation in reality, and for many years ceased even to read a newspaper regularly.[2] Dr Haycraft, in that interesting forerunner of eugenical literature proper, *Darwinism and Race Progress*, sounded what was to be the familiar note, when he wrote: 'what are the petty combinations of parties, or even those temporary associations of individuals, which aim at a common or national policy, by the side of the health and the capacity of that race of which we are but passing representatives . . . ?[3]

That eugenics would sooner or later engage the attentions of statesmen and government was taken for granted. But eugenists regretfully concluded that politicians were a timid group of people, who shied away from anything unusual or controversial. In Tredgold's words,

> Under a democratic form of government no legislation much ahead of public opinion can be carried; and this is the chief danger arising when a democratic form of government is evolved in advance of the education of the democracy. In such circumstances the combination of an imperfectly informed electorate with a paid professional legislature is only too apt to conduce to the establishment of a vicious circle, in which true social science is prostituted by the promulgation of so-called reforms which are a mere pandering to the present, rather than part of a definite system designed to further the real development and progress of the nation.[4]

This was the reason why Dean Inge thought democracy was 'perhaps the silliest of all fetishes that are seriously worshipped among us . . .'[5] Arnold White similarly argued that 'Democratic Government means the inclusion of those who are ignorant of the laws of hereditary transmission, who are least prepared to lay down present ease for future good, and who are least accustomed to resist the impulse of passion or the suggestions of desire'.[6] This problem was thought by Pearson to be well-nigh insuperable: 'a government which drew a line between capable and incapable would rapidly perish: for the incapables care nothing for the future of the race or nation, but seek from their necessarily subservient governments "panem et circenses"—more time to pillion-ride, more leisure for cigarettes, chocolates and cinemas—at the cost of the capable'. How, in these circumstances, could a start be made in 'practical Eugenics'? 'We might as successfully ask the weeds in a garden to make way of their own accord for the flowering plants whose development they choke. Let my readers think what a gardener could achieve, if his tenure of office depended on the consent of the weeds!'[7]

Inevitably, then, many eugenists were carried away by the logic of their argument to advocate some authoritarian form of government in which 'experts' could attend to racial issues, undistracted by the clamour of the mob. Others, like Miss Elderton, appealed desperately for some statesman to come forward and give a lead while there was yet time.[8] Whetham, on the other hand, made the characteristic suggestion that eugenical problems might be handled more sensibly by the House of Lords, a body which did not have the fear of the polls before its eyes and so was less likely to 'waste' its time on pointless political controversies.[9]

The general view of the politician was thus, to quote Dr Slaughter, that he was 'merely a puppet of public opinion whose greatest desire [was] to keep his place as long as possible while doing that minimum of work which [was] forced upon him'.[10] But eugenists grumbled that members of Parliament did not even seem to be aware of their own inadequacies. For this they had a simple explanation: 'It is to be feared that not many of our parliamentarians have had any training in biology'.[11] Bateson could never get over his shock on first discovering that Mr Gladstone thought that human beings normally had twenty-eight teeth: 'Portentous ignorance of this kind is common among historians and legislators. In itself perhaps a trifle, it is a symptom of detachment from the actual world so complete as to disqualify a man from safely exercising high functions of statesmanship, demanding, as they must, a discernment which can only come from wide knowledge of natural fact'.[12]

Pearson indicted liberalism on the self-same count: 'it is the ignorance of great biological truths which has been the blot on the Liberalism of the past; it is this ignorance which makes so much of Liberal social effort vain. You cannot, sir, reform man, until you understand the factors which control his growth, and you cannot understand those factors by endlessly talking about them . . .'[13] This impatience with argument and assertion, which lacked scientific backing, was one of Pearson's favourite hobby-horses. Sloppy thinking and dangerous, because misdirected, humanitarian sentiment would eventually disappear when a new generation of statesmen-philanthropists and civil servants had emerged, equipped with a proper statistical training.[14] Already, he believed, intelligent working men, sickened by 'rhetoric and verbal controversy', were manifesting a marked lack of confidence in the leaders of both political parties, a circumstance which he used to explain the low turn-out in the 1910 Elections.[15] Sooner or later, he said, it would be more generally understood that the great social problems of the day could no more be solved by consulting popular opinion than could questions of astronomy and physics; these were difficult problems requiring special training and analysis of a sort wholly beyond most people's grasp; meanwhile, party methods and hustings oratory merely created confusion and prevented the formulation of a scientifically-based policy that would lead to the consequence intended by its authors.[16]

Eugenists, therefore, fell into the unfortunate habit of damning politicians up hill and down dale. Sir James Barr, a Vice-President of the E.E.S., was especially prone to do this, and his gibes were all the more ill-chosen in that he found it impossible to conceal his hatred of the Asquith Administration.[17] But even Saleeby, whose sympathies probably lay with the Radical Party, and who was certainly not a Conservative, joined in the game of spattering the professional politicians with insults. 'The eugenist has nothing to do with the low game called party politics', he wrote. A great and growing section of the community had come to see 'party-politics for the humbug and sham that it is, and the House of Commons as a lethal chamber for souls'.[18] On other occasions he hit out against 'glass-eyed politicians' wrangling in 'the House of Gramophones', and repeated with relish Adam Smith's dismissal of 'that invidious and crafty animal vulgarly called a statesman or politician'.[19]

When, in 1925, Schiller looked at the political parties to see which of them could most easily be brought to listen to reason, he found that they all entertained prejudices unfavourable to eugenics.

> The Conservatives may be supposed to have most natural sympathy with the aim of arresting the elimination of the best; but they are no longer the aristocrats they were, and the party is falling more and more under the influence of industrial potentates greatly interested in promoting the abundance of cheap

labour. Also it cannot be denied that the idea of eugenics is new, and therefore suspect. The Liberals, on the other hand, though not hostile to change as such, are not specially favourable to science, and are tainted with a false humanitarianism which aggravates, and does not cure, social maladjustments; while the Labour Party, though it ought to be the most reluctant to work for the support of wastrels and parasites, has unfortunately got into the way of regarding limitation of output as a legitimate way of raising the social value of any product.

Schiller ended with the despairing conclusion that it might be necessary to abandon all the existing organizations, and 'found a new party of eugenical reform'.[20] But, like his earlier hope that the 'old-fashioned politicians' might undergo a conversion, publicly confess the errors of their ways, and embrace the cause of Social Hygiene,[21] this suggestion was lacking in common sense. A more realistic position was that of Saleeby, who told an interviewer from the *Daily Sketch* after the January 1910 Elections that he was not surprised at the absence from the Commons of avowed race-culturists, but dismissed the idea of running eugenics candidates: 'We shall have to work through the great historic parties. We must get a Parliament returned pledged to deal with the whole question of inherited disease, the feeble-minded, inebriates, and the hereditary unemployable'.[22]

In the election that had just been held, the E.E.S. had, in fact, started, albeit tentatively, to act as a pressure group that could extract pledges from candidates. On 12 January 1910 it was decided, at the last moment, to circularize all former Members whose private addresses were readily available, on the eugenic aspects of Poor Law reform, and ask them: 'Would you undertake to support measures recommended in the Report of the Poor Law that tend to discourage parenthood on the part of the Feeble-Minded and other degenerate types'. Over one hundred letters were sent out, and many favourable replies received.[23] The campaign for the Mental Deficiency Act organized by the E.E.S. over the next three years was an example of the success that could be obtained by a patient cultivation of M.P.s on an issue where legislation for a particular objective was a practical possibility. But Parliamentary propaganda as a whole was allowed to languish. In March 1910, the Council of the E.E.S. resolved 'that in the present state of political chaos it was quite useless to attempt propagandist work among Members of Parliament, so the matter was left in abeyance'.[24] In November 1911, a Parliamentary Committee was set up for the first time, with the task of watching all bills going through Parliament that were of concern to eugenists.[25] But one of its members, Mr Marshall, was complaining in July 1914 that the activities of the Society in this matter were still quite inadequate, and instanced the recent Budget Debate where the eugenic point of view on child rebates had not been seriously pressed. Further organizational reforms were under way when the First World War broke out.[26]

Eugenists were undoubtedly hampered in their effectiveness as a pressure group by the way they treated politicians as a group. Little wonder that, subjected to a constant stream of abuse, politicians kept aloof from the E.E.S. Although it was making considerable headway in 1913, and enjoyed a good deal of sympathy, if not active support, the Society recruited a mere handful of M.P.s in three years. Balfour's acceptance of Honorary Membership was an isolated triumph; a similar offer to Asquith in 1913 met with a predictable refusal.[27] Eugenists paid the penalty of indecision; they could not decide whether they wished to enter the political arena and exercise there what influence they could, or whether their aim was the conversion of the thinking public to a new outlook on life which would make all existing political philosophies and organizations superfluous. Karl Pearson, for one, clung to the whimsical idea that eugenics, as he presented it to the world, was politically neutral, a mere summary of the findings of science. This conviction led to intransigence and arrogance; politicians were hectored and scolded for their ignorance, stupidity, and immoral craving for electoral popularity. In this way opponents were made of many politicians who basically shared the fears and hopes which the eugenics programme expressed.

At this point it might well be asked: what, if anything, did eugenists want government to do for them? What precisely would they themselves have done if by chance they had attained power? In a nutshell, the eugenic demand was that steps should be taken to stimulate the fertility of the 'best stocks' and to reduce the fertility of the 'worst'. Saleeby coined the terms 'positive' and 'negative eugenics' to denote this distinction between the two aspects of the subject and since these terms quickly became part of the accepted language of eugenics they may as well be used here.

7

Positive Eugenics

In his Huxley Lecture of 1901, Galton had quite explicitly declared: 'The possibility of improving the race of a nation depends on the power of increasing the productivity of the best stock. This is far more important than that of repressing the productivity of the worst . . .'[1] Even at the end of his life, while admitting that the Report of the Royal Commission on the Feeble-Minded had made 'negative eugenics' quite the most 'pressing' task facing the eugenist,[2] he never came to share the interest of his followers in diseased and pathological families and cases of 'morbid inheritance'. Galton continued rather to work along the lines established in his earlier books, *Hereditary Genius* and *English Men of Science*, in which he examined families containing a large number of exceptionally gifted individuals.

The hope that at some future time the human race might be largely composed of men and women possessing the illustrious qualities of Shakespeare and Darwin was a continuing source of inspiration to eugenists. After quoting Nietzsche on the 'Superman', R. A. Fisher told the Cambridge University Eugenics Society: 'We can set no limit to human potentialities, all that is best in man can be bettered; it is not a question of producing a highly efficient machine, or a paragon of the negative virtues, but of quickening all the distinctively human features, all that is best in man, all the different qualities, some obvious, some infinitely subtle, which we recognise as humanly excellent'.[3] Bernard Shaw was fond of asking what hope existed for the species when it was clear that there had been no advance on Socrates and Plato. Even Beatrice Webb, after seeing a performance of Shaw's *Man and Superman*, was moved to write in her diaries:

'*We* cannot touch the subject of human breeding—it is not ripe for the mere industry of induction, and yet I realise that it is the most important of all questions, this breeding of the right sort of man . . .'[4]

Breeding experiments were never, of course, envisaged by any responsible eugenist. Galton himself feared that in man's present state of ignorance, 'he might possibly even do more harm than good to the race' by attempting to arrange eugenic marriages.[5] Moreover, as Major Darwin admitted, 'the paramount necessity of maintaining a moral code introduces vast difficulties in the case of man which are unknown in the stock yard, and unquestionably the possibilities open to us are thus greatly limited'.[6] The eugenists would have been less liable to misconstruction if they themselves had not so frequently drawn analogies between human procreation and stock breeding. Essentially, however, they were doing no more than pointing out that the human race was a part of the organic world and thus, from the biological point of view, subject to the same laws of selection and variation. This was also what Pearson meant when he wrote that 'were we the "superman" we could breed a race of abnormally shy men, as we could breed a race of abnormally tall men; and we could breed a race in which six fingers were the rule or one in which nearly every members was a deaf-mute'.[7] Not for one moment was he suggesting that such crazy experiments should be mounted.

Galton perfectly well understood that, apart from the obvious ethical objections against 'breeding for points', it was ruled out of account by the variegated character of any complex human society, which required for its efficient functioning very different kinds of skill and personality, all contributing to the fullness and richness of life. In deciding on how best the breed could be improved, Galton therefore suggested that special aptitudes would have to be assessed by those who possessed them. 'Thus the worth of soldiers would be such as it would be rated by respected soldiers, students by students, business men by business men, artists by artists, and so on'. Criminals, however, were not to be permitted to form their own associations, with similar voting rights, nor were other occupational groups which society deemed undesirable.[8] This scheme, however, had obvious limitations. It was also controversial in that it assumed both the continuance and the legitimacy of the existing social order. There are many other examples of Galton's habit of treating classes and occupational groupings as something fixed and immutable.

But Galton also believed that there was, beside professional aptitude, a more general quality, or rather a complex of qualities, which society could and should promote. He described what he meant by the phrases 'civic worth' or 'worth'. By this he hoped to describe those human beings who possessed an above-average share of 'at least goodness of constitution, of physique, and of mental capacity'.[9] If this was somewhat vague, Galton's followers were unable to improve much on it. Crackanthorpe suggested that 'sound health, a sufficient amount of energy, a well-balanced brain [were] obviously desirable, nay, necessary ingredients'.[10] Robert Ewart defined the excellence for which eugenists were striving in terms of height, scholastic aptitude, and power of survival.[11]

It will be noted that Crackanthorpe and Ewart both followed Galton's lead in giving priority to the physical attributes. This was not because Galton did not value man's spiritual and intellectual qualities. However, he believed that there was a very close correlation between all these attributes, and physique had the advantage of being more precisely and objectively measurable than, say, intelligence. 'The youths who became judges, bishops, statesmen, and leaders of progress in England', he asserted, 'could have furnished formidable athletic teams in their time. There is a tale, I know not how far founded on fact, that Queen Elizabeth had an eye to the calves of the legs of those she selected for bishops. There is something to be said in favour of selecting men by their physical characteristics for other than physical purposes. It would decidedly be safer to do so than to trust to pure chance.'[12] This belief in a correlation between physical health and intellectual ability was held by many eugenists, including Ewart and the Whethams.[13] In their understandable irritation against the fashionable notion that 'genius' was a pathological condition, a form of abnormality, often going hand in hand with insanity and disease, eugenists often went to the opposite extreme; thus, James Barr, on the subject of the 'consumptive poet', said: 'Robert Louis Stevenson was a beautiful writer, and many of his epigrams are very fine, but much of his writings will not bear analysis according to the hard rules of facts, and I am convinced that if he had not been phthisical he would have written better and much more sanely'.[14] If eugenists wanted to point to a living embodiment of their ideal of the *mens sana in corpore sano* they had no need to look further than Galton himself, who provided an especially apt example in that he came from precisely the kind of 'eugenic' family he had taken pains to investigate.[15]

Unfortunately, Galton's belief that physical health and strength were an outward and visible sign of man's inward grace was soon to take a hard knock at the hands of his own biometric disciples. The statistical investigations undertaken by Karl Pearson on material furnished by L.C.C. schoolteachers suggested that, although both physical and 'psychical' characters were inherited to a marked degree, the correlation *between* them was not high. As he put it in a latter summary of his researches: 'Taking an all-round view of the matter, it is surprising how small is the relationship of general health to the psychical characters. It would seem that far too much weight has been attributed to the health factor in dealing with mental characteristics'. In retrospect, Pearson felt that this exaggerated view of the importance of health had come about because of an excessive concentration on poor children in elementary schools, whose poor physique was in many cases the result of defective environment, want of medical care, and even deficiency of food. In the more normal state of affairs encountered among middle-class schoolchildren, these 'interfering factors' were absent.[16] Pearson, nevertheless, continued to set great store by physical health and vigour, which he thought should be considered, along with the more narrowly intellectual qualities, when selections were being made for secondary schools and universities. The picture he attempted to conjure up was one of weedy scholarship boys entering institutions of higher education when they lacked the physical stamina (and perhaps the moral character) to benefit from the instruction they were about to receive.[17] Pearson's advocacy of a 'muscular intellectuality' throws an interesting light on the values cherished by many eugenists of his day. But it matters very little compared with his damaging admission that health and intelligence were not closely correlated.

For this went a long way to undermining Galton's arguments about 'civic worth'. And back once more came the old conundrum about what eugenists were going to do about the consumptive poet and the sickly 'genius'. Galton, who had not seen this as a serious problem at all, had suggested that inferiority under any of the three heads of ability, physique, and character should outweigh superiority in the other two.[18] But it was easy to compile a long list of 'famous men' who would have failed to pass this eugenic test; and opponents of the movement had an enjoyable time in pointing out that eugenists might have 'predicted' the ill-health of a Carlyle, Ruskin, Keats, or Stevenson, without also predicting their remarkable qualities. Major Darwin could only get out of this fix by suggesting that 'perhaps marked excellence in any desirable quality ought to go very far towards outweighing any undesirable qualities', an inversion of Galton's scale of priorities.[19] But Bateson was not alone in wondering whether 'by the exercise of continuous eugenic caution the world might have lost Beethoven and Keats, perhaps even Francis Bacon', and whether 'civic worth' might not exclude the 'unbusinesslike Bohemians', artists, musicians, authors, discoverers, and inventors, 'literally the salt of the earth'. 'Civic worth', he warned, might become a mere synonym for mediocrity, dullness and conformity.[20]

It is clear in retrospect that the eugenists should have left the question of 'genius' strictly alone. To talk as though they had some blueprint for creating new Shakespeares and Newtons was to invite ridicule. Professor Lindsay, who admitted that genius had 'a somewhat sorry record from the strictly Eugenist point of view', claimed that eugenics was essentially concerned 'with the production of talent, ability . . . the qualities which make a man or a woman a serviceable social unit'.[21] There were, perhaps, three reasons why 'genius' should have so frequently been discussed. Firstly, the critics of the movement and innocent enquirers after truth persisted in raising the issue. Secondly, the title of Galton's first major work, *Hereditary Genius*, gave a misleading idea of what he and his followers were setting out to do. Thirdly, there was so much loose talk current in the immediate pre-war years about the superman, that half-understood Nietzschean concepts became entangled in the eugenics debate, though few serious eugenists desired this.[22] An example of the way in which the New Biology and Nietzsche were fused is provided by Eder's articles on eugenics in the *New Age* in 1908, in which the superman was described as a mutation or giant variation comparable to the mutations discovered by de Vries in his work on the evening primrose.[23] Shaw added to the confusion by linking Nietzsche to Lamarck's evolutionary views, which no self-respecting eugenist and few biologists now thought had any validity.

However, Lindsay's suggestion that eugenics, far from having anything to do with these utopian speculations, was about fostering those qualities which made men and women 'serviceable social units' raised fresh difficulties of its own. It drew attention to Galton's indecision about whether specialized aptitude or 'all-round ability', social usefulness or a versatile individuality, were the more important to the future of the race. Most eugenists, including Galton, favoured general excellence and individualism; and none did so with such fervour as 'reform eugenists', like Lowes Dickinson, who thought that a 'high average of capacity' was a necessary concomitant of democracy.[24] It was this emphasis on individuality which goaded Benjamin Kidd into protesting that the struggle for existence would select that nation which had the highest '*social* efficiency', something that was different from, and even opposed to, individual efficiency.[25] Similarly, Chiozza Money felt obliged to put in a kind word for the 'decent duffer'.[26] Money obviously did not share Kidd's political philosophy, but both men were agreed that it made no sense to discuss individual excellence without reference to the society of which these individuals formed a part.

It was precisely here that Galton's conception of 'civic worth' was most vulnerable to criticism. It defined particular social or moral qualities as worthy of propagation, without supplying a political philosophy that justified these preferences. Indeed, Galton once conceded that 'the goodness or badness of character is not absolute, but relative to the current form of civilization'.[27] This was precisely the problem which he should have confronted head-on. Instead he largely evaded it. One disillusioned eugenist complained of this in a letter to *The Times*; sound organization, he argued, rested on ideas, ideals, and beliefs, but 'Sir Francis Galton in the paper in which he first outlined his new science of race improvement, proposed to leave morals out of account on the ground that it was too complex a subject. The truth, I am afraid, is that the eugenist so far has mistaken the problem. He is dealing at present only with a problem of individuals and not with the problems of society and race improvement'.[28]

L. T. Hobhouse made the same point when he defined Progress as 'the realization of an ethical order'; but, on their own admission, eugenists had no clear idea of what they meant by 'social worth'; only philosophy, he argued, could supply the necessary ethical goals, which biology alone was powerless to do.[29] Urwick had some fun by pointing out the disagreements between eugenists over what human attributes were to be most highly esteemed. More seriously, he observed that 'society . . . is always trying to get people "good" and to get good people,—without ever knowing clearly what it means by "good"; for it means something far too complex to be defined, and something, too, which changes as the social ideals change'; eugenics was distinctive in defining goodness quite simply in terms of certain physical and mental qualities and prescribing precise methods for attaining it; but society would never accept this restriction of its ethical outlook.[30] In vain did the Oxford philosopher, Schiller, attempt to defend eugenics from criticisms of this kind; no-one, said Schiller, proposed to carry out a eugenical *coup d'état*; being aware of their liability to err, the proponents of eugenics would guard against this danger by incorporating into their actions provision for detecting and correcting any possible errors; the ideals of the eugenist, like the ideals of other people, were not 'prior to the particular experiences they profess to "explain", but are built up out of suggestions derived from the latter'.[31] Schiller's line of reasoning, unfortunately, was too sophisticated for most eugenists to follow, and even if it might have removed some of the animosities which eugenics aroused, it also blunted its attractiveness.

Finally, leaving aside the ethical question for the moment, there was the objection, which Galton himself felt to have some force, that knowledge about the mechanism of heredity had not advanced anything like far enough for scientists to be able to predict the probable characteristics of the offspring of a given pair of parents. Despite Pearson's rhetorical flourishes, it seemed improbable that a race of 'shy men' could have been bred, though a race of 'exceptionally tall men' would have presented fewer difficulties. The *Nation* caricatured 'Positive Eugenics' as the belief that race progress could be achieved by 'selecting the good, and breeding Beethovens or

Shakespeares as a pigeon fancier breeds fantails'. The writer concluded that 'if we suppose adequate knowledge of all we wish to breed for—that is to say, of all the elements of real value to a society on the one hand, or if we grant further a power existing in society impartial enough to apply this knowledge without fear or favour, and strong enough to impose acceptance on individual men and women, we should admit that eugenics would become a very important branch of applied social science'. But at present all this was mere wishful thinking. 'Until the biologists have settled among themselves what heredity is and how it works', wrote the *Nation*, 'they cannot intervene with much effect outside their own province'.[32]

Moreover, the qualities that were generally admired in 'great men' were often far too complex to be easily *comprehended*, still less to be scientifically produced. Hobhouse suggested that 'it is quite possible . . . that two strains, each sound in itself, should when united produce a bad result', and that, conversely, 'some stocks desirable in themselves contain strains that suitably blended with others are of value to the rational character as a whole'.[33] H. G. Wells, like Hobhouse, stressed that even those qualities which most people thought to be worth cultivating amongst human beings, health, energy, beauty, were not simple qualities as at first sight they might appear to be, but a harmonious balance of elements that arranged in a somewhat different manner would produce quite a different result. These were not 'points' that the scientist could breed for.[34] Finally, there was the problem, which many eugenists rather glossed over, of social circumstance. Galton singled out for especial praise the quality of energy, but the very impulses which brought one man eminence might lead another, less fortunately circumstanced, to gaol.

If all these difficulties were taken into account, what was there left to be salvaged from 'positive eugenics'? In the eyes of most of Galton's followers, comparatively little. When Saleeby showed Galton an article on eugenics shortly to be published in the *Sociological Review*, the latter commented that the section on 'positive Eugenics' was very short; 'so, indeed, it was', observed Saleeby, 'but at that date we had no knowledge of the exact inheritance of valuable qualities'.[35] Thus, a common view among eugenists was that again expressed by Saleeby: 'we may not know

what worth is, but we can all recognize "unworth"';[36] hence, it was better to concentrate on 'negative eugenics'.

But this new concern with pathological stocks necessarily involved eugenists in attempting to capture political power or at least influence governments, something which Galton himself had never seriously intended. Galton's hope was that of winning over strategically placed individuals, so as to alter the state of public opinion on questions of marriage, reproduction and race-improvement. His lack of interest in the political process itself is revealed in the one and only address which he made to the E.E.S. in October 1908. Already local associations were coming into existence, affiliated to the central body. Galton, therefore, mused aloud before this audience about what functions these new organizations could most usefully serve. In brief, his proposal was that men of standing and importance in the local community, like doctors, clergymen, Medical Officers of Health, and lawyers, should be persuaded to join together in social gatherings where knowledge of heredity could be disseminated, pedigrees of local families of note compiled, and means found of enabling promising young people to make their start in life. Members of the local associations would probably be those who considered themselves or were considered by others to be 'the possessors of notable eugenic qualities'.[82] Galton added: 'It ought not to be difficult to arouse in the inhabitants a just pride in their own civic worthiness, analogous to the pride which a soldier feels in the good reputation of his regiment or a lad in that of his school. By this means a strong local eugenic opinion might easily be formed . . .'[37]

Galton's audience listened, as always, with respect to their patron's views, and then proceeded to ignore them. The local branches of the E.E.S. did, along with the central organization, go in for some desultory drawing up of pedigrees. Occasionally, an article appeared in the *Eugenics Review*, such as J. F. Tocher's 'The Necessity for a National Eugenic Survey', of July 1910, with its odd proposal for eugenic clans and septs, which attempted to extend Galton's line of reasoning.[38] But far from forming social clubs of self-conscious, practicing eugenes, these branches soon took on the role of local propaganda organizations, concerned with the same sort of polemical and political work that largely preoccupied the parent body in London. No doubt many of the men and women drawn into the movement did take a complacent view of their own genetic endowment. On rare occasions the membership was even exhorted from the platform to see itself as the forerunners of the superior species which scientists could now predict. R. A. Fisher apostrophised the Cambridge University audience who attended his lecture in November 1912 as 'agents of a new phase of evolution', who had the mission of spreading abroad, 'not by precept only, but by example, the doctrine of a new natural ability of worth and blood'.[39] But this note was not often struck.

Neither was the emphasis that Galton placed upon private philanthropy a characteristic of the eugenics movement as a whole. Galton, as we have seen, has been sometimes criticized, and not without reason, for identifying eugenic classes with social classes. Nevertheless, he was not so naïve as to suppose that economic status by itself denoted civil worth. In fact, he was disturbed by the discrepancy that often appeared between the two conditions. Hence, Galton devised ingenious and detailed plans for providing financial incentives to 'eugenic' couples to marry early in life, so that they might be permitted to start having a family while the wife still had many years of child-bearing ahead of her. 'The means that might be employed to compass these ends', wrote Galton, 'are dowries, especially for those to whom moderate sums are important, assured help in emergencies during the early years of married life, healthy homes, the pressure of public opinion, honours, and above all else the introduction of motives of religous or quasi-religious character.' He also sketched out a plan for 'the provision to exceptionally promising young couples of healthy and convenient houses at low rentals', which might develop into settlements complete with self-selected fellowships and rooms of residence, with something of the atmosphere of a socially prestigious club: 'The tone of the place would be higher than elsewhere, on account of the high quality of the inmates, and it would be distinguished by an air of energy, intelligence, health and self-respect and by mutual helpfulness'.[40] That this imaginary set-up derives from a rather rosy view of Oxford and Cambridge seems highly probable, and is confirmed by Galton's unpublished utopian novel, 'Kantsaywhere', in which the analogies between a eugenic community and a university college are explicitly drawn.[41]

Who would be selected for this favoured treatment? Galton allots the task of selecting young men of promise to an independent committee connected with the universities, which would have evidence about the candidates' athletic ability as well as their academic performance; there would, in addition, be personal interviews and a study of the family history of all the candidates. The social groups upon which Galton hoped to draw for his 'Eugenes' were obviously somewhat restricted; he seems to have been thinking mainly of young people from 'good' families whose members had, through no fault of their own and through no want of merit, fallen on hard times.[42] The same impression is conveyed by his advocacy of a rejuvenated patronage system, 'a kindly and honourable relation between a wealthy man who had made his position in the world and a youth who is avowedly his equal in natural gifts, but who has yet to make it'. In a naïve analogy, Galton spoke of noble families getting 'fine specimens of humanity around them', rather as they 'procure and maintain fine breeds of cattle and so forth, which are costly, but repay in satisfaction'.[43] Putting the trust that he did in enlightened eugenic action by the well-to-do, he had little, however, to say about the role of State aid and State legislation, matters which engrossed a later generation of eugenists, many of whom must have thought, though they were too polite to say so, that Galton was blinkered by his mid-Victorian political philosophy, and was merely proposing to tinker with the forces that threatened national decay.

What Galton really wanted was for there to be continuous enquiries into the facts of heredity, together with the construction of biographies and pedigrees which would provide 'an exact stocktaking of our worth as a nation'. Once the implication of these researches had sunk in, he seems to have supposed, much else would follow as a matter of course. Ties of friendship would grow up among the gifted families; they would tend to intermarry, and in time there would be 'a "golden book" of natural nobility'.[44] Whetham later took up this last suggestion and envisaged a future 'when an aristocracy of Eugenic families [would] arise, who [would] guard their inborn inheritance as carefully as ever the older aristocracies [had] kept the purity of blood which could boast of *Seize Quartiers*'.[45]

In the hands of the 'reform eugenists' this emphasis upon the desirability of replacing an artificial nobility, based upon wealth and tradition, by a natural nobility whose qualities were inborn, could be used as a way of attacking the rottenness of contemporary society. The Whethams by no means intended to create this effect. Nor did the conservative Galton, who, although anxious to create a new 'sentiment of caste among those who [were] naturally gifted', said he did not propose 'to begin by breaking up old feelings of social status, but to build up a caste *within* each of the groups into which rank, wealth and pursuits already divide society, mankind being quite numerous enough to admit of this sub-classification'.[46]

Nevertheless, Galton admitted that in a society dominated by a 'false' theory of democracy, the members of the eugenic caste, even if scattered throughout society, might attract jealousy and persecution; in self-defence they might decide to form colonies of their own, engage in co-operative enterprises of various kinds, and lead an exclusive self-contained life; if they were still exposed to harassment, wrote Galton, 'let them take ship and emigrate and become the parents of a new state, with a glorious future'.[47] Over thirty years after penning these words, Galton allowed his mind to play over the question of how such a Eugenic State might be organized, in his projected book, 'Kantsaywhere'.

Similar ideas were quite widespread around the turn of the century. In 1891, Charles Wicksteed Armstrong published a novel depicting a eugenic community, and in the 1920s he returned to the subject of creating a settlement, with its location, perhaps, in the South-Eastern Pyrenees, to be peopled by men and women whose 'worth' had been attested by an expert 'council of management'.[48] This scheme remained as visionary as the fictional 'Arlington Community', described by the anonymous author of the *Essays in Buff* (1903): a community in which controlled breeding experiments were conducted, rather along the lines of those which John Noyes had once supervised in the Oneida Settlement.[49] But was it sensible for eugenists even to *aspire* to forming colonies of their own? Major Darwin thought not. Apart from the probability of 'regression' over a number of generations, such a community, he suggested, would arouse fierce resentment among the excluded, and it would soon face the threat of annihilation at the hands of the 'unfit'.[50] In any case, eugenists were, as a body, now committed to seeking reforms within existing society, not withdrawing from it into exclusive little sects. If the human race were to be saved from deterioration through the multiplication of the unfit, withdrawal from the mainstream of society was not the best way to arrest this decay.

The only concrete recommendations in the sphere of positive eugenics to find much support were marriage certificates and financial bonuses from the State to parents of fit offspring. Galton himself had proposed that at some future time it might prove practical to start a formal system of 'eugenic certificates', which could be granted by an appropriate authority to those applicants who could prove that they had at least an average share of good health, physique, and mental capacity.[51] Havelock Ellis was quick to see that here was a possible device for encouraging eugenic marriages. No one would be compelled to submit himself to an examination in the hope of receiving one of these certificates, argued Ellis, just as no one was compelled to seek a university degree; but its possession would greatly increase a man's or woman's chances of making a 'good' marriage.[52] Dr Slaughter's thoughts were running along the same lines: 'As courtship is a process of suggesting or displaying qualities and possessions, it may be that part of its regular routine will be the exhibition of the life insurance policy; fond parents should be at least as interested in this as in the young man's actual or prospective balance at the bank'.[53] On Galton's suggestion, Ellis actually tried to work out a practical scheme for the issuing of eugenic certificates; Galton was disappointed at how costly it would prove to administer.[54] Perhaps for this reason, the idea subsequently fell somewhat into the background. Although marriage certificates often feature in eugenical literature, they usually appear within the context of 'negative eugenics', being presented rather as a means of discouraging *unsuitable* marriages than as a device for stimulating the fertility of the 'fit'.

In any case, the problem was not so much one of engineering 'suitable' marriages, but one of encouraging fit parents to go back to the large families of their grandparents' days. This brings one logically on to a consideration of what was currently called, 'the endowment of motherhood', that is to say, family allowances. The Fabian Society was busily engaged in promoting this reform, and some eugenists gave it their support. For example, Dr Slaughter is reported as saying on one occasion that working women might have to be paid by the State during the time of child rearing and be given a sum to provide towards the child's upbringing,[55] and even the Whethams once supported this proposal, at least as an interim measure.[56]

But a more common response was that of Montague Crackanthorpe, who, noting that the Webb's endowment of motherhood scheme made no distinction between 'good' and 'bad' parentage, commented: 'The old, old story, this, of quantity versus quality'.[57] Professor Lindsay was similarly sceptical about whether the eugenist should endorse the 'endowment of motherhood': 'Not motherhood as such, but good motherhood, is his ideal. Large families may be either a boon or a burden to the State, according to the quality of the offspring. The subsidising of marriage without due precautions, might tend to accentuate existing evils'.[58] Lowes Dickinson might believe that public opinion would soon be ready to accept a system of State payments for children only in the case of parents whose union was approved of by society.[59] But most eugenists, including those with a progressive social outlook, like Sidney Herbert, thought that such a discrimination was a practical impossibility. It was also Herbert who noted that 'the very classes who are lowest in the economic scale, and are most in need of such aid, have already the highest birth-rate'.[60] Most eugenists were aghast at the prospect of still further stimulating the fertility of those social groups who, in their eyes, constituted the 'unfit'.

Refinements could be made to the Webbs' proposals.[61] But there would still be the sceptics who doubted (and with good reason) whether financial inducements of a direct kind would ever achieve their avowed purpose. The Whethams, for example, pointed to France, where the State, the municipalities, and some big commercial concerns were all hoping to stimulate the birth rate in this way, so far without any observable effect.[62] From the 'conservative' wing of the movement came two further attacks on the kind of 'endowment of motherhood' scheme favoured by the Webbs. Firstly, such a piece of 'meddling socialism', they said, would be extremely costly; already, according to one contributor to the *Eugenics Review*, the feeding of necessitous schoolchildren was putting 'a heavy and increasing burden' on the ratepayer: 'To impose on him the further burden of an "Endowment of Motherhood" rate or tax would well-nigh break his back'.[63] Secondly, there was the argument that a fixed bonus for every child, ignoring differences of living standards in working and middle class families, would be 'unfair'; 'the only sound principle', wrote the Whethams, 'would be to make for each child an allowance which rose in proportion to the income of the parents till a limit was reached depending on the amount reasonably to be spent on the maintenance and education of children in the upper classes'.[64] This proposal anticipated the position adopted by the E.E.S. after the First World War. But the perpetuation and strengthening of class divisions and income differentials was, of course, the very opposite of what the Fabian proponents of the endowment of motherhood hoped to achieve.

Rather surprisingly, even Saleeby declared himself 'opposed to the principle of bribing a woman to become a mother, whether overtly or covertly, whether in the guise of State-aid or in the form of eugenic premiums for maternity . . . I do not see what service it renders to motherhood in its psychical aspects'.[65] He added, 'the so-called endowment of motherhood, by the State, proposes to serve motherhood by discharging fatherhood from its duties. On whatever round the feet of Progress and Eugenics may fare, this is none of them. It is not progress, but full retreat, helter-skelter back to the beast'. But Saleeby had no objection to the provisions for maternity benefit in Lloyd George's National Insurance Scheme, since this compelled the *father* to set aside money for the care of his wife when she became pregnant.[66] Moreover, said Saleeby, the government should at least stop penalizing marriage. Eugenists instanced the now defunct celibacy regulations of Oxford and Cambridge Colleges as the sort of folly that must at all costs be avoided in the future.[67]

On the positive side, Karl Pearson suggested that the State might follow the example of the Indian Civil Service and take parenthood into account in the determination of the salaries of public employees.[68] William McDougall had been one of the pioneers of this idea, suggesting in a contribution to the Sociological Society in 1906 that the salaries of higher civil servants could be adjusted according to the number of their children, and perhaps a similar system applied to certain grades of inspector, local government officials, and occupants of university chairs.[69] This might still have seemed to Saleeby an objectionable sort of 'bribe'. But at least it avoided the danger of encouraging an *indiscriminate* rise in the birth-rate. The same is true of Pearson's vague advocacy of a system of insurance in which employer, State, and workmen would combine to insure against invalidity, motherhood, and the nurture of offspring—all provision being differentiated by the fitness of the parentage.[70]

The most favoured expedient, however, was an alteration of the tax system, so as to discriminate in favour of married couples with dependent children. The beauty of this idea, in the eyes of most eugenists, was that it could not possibly encourage the multiplication of the unfit, since, broadly speaking, only the middle classes paid income tax. The Liberal Government, for reasons of its own, began to move in the direction favoured by eugenists when, in the People's Budget, it introduced an allowance of £10 a year for every child below sixteen years of age in the case of income tax payers whose income fell below £500 a year. In the 1914 Budget, Lloyd George doubled the rebate (popularly called, 'the Brat'). This not only delighted the Webbs, who hailed this action as the first timid government recognition of the desirability of the endowment of motherhood,[71] but it also encouraged the eugenists proper; Karl Pearson, for example, observed, 'It will be the fault of eugenic workers if the thin edge of the wedge thus inserted be not driven home'.[72]

The only complaint of the E.E.S. was that Lloyd George's concessions did not go far enough. In the spring of 1914, the Society lobbied sympathetic M.P.s, and tried without success to inspire amendments to the Finance Bill. Eugenists were able to do this since, for once, they had previously worked out something that could be called an agreed policy on the whole issue of child rebates. William Marshall, the leading spirit in the Haslemere branch of the E.E.S., had persuaded the Council of the Society, in 1911, to hold an inquiry into the issue.[73] By 1914, Darwin had worked out a eugenic scheme; the substance of it can be discerned from a memorandum, which survives in the Arnold White Papers,[74] but it can be most conveniently studied in the evidence submitted by Darwin to the Royal Commission on the Income Tax in October 1919.[75]

As an interim measure, Darwin wanted to ease the burden of parenthood by allowing income-tax payers a rebate which would partly offset the cost of educating their children. Another way of giving relief to the professional man with a large family, he argued, would be to make the fixed child rebate allowance payable to people earning more than the existing maximum (which had risen to £1,000 by 1919). This last suggestion had been brusquely rejected by Lloyd George in 1914,[76] but the Royal Commission caused pleasure among eugenists by recommending the extension of child abatements to all income-tax payers, whatever their income.[77] But the most radical of Darwin's suggestions was that the income of a family should 'count as a number of separate incomes, the number being equal to or dependent on the number in the family'. He insisted that this drastic reform could be carried out in such a way that each economic stratum would continue to make the same contribution to the revenue of the State, although within the income-tax paying class the burden would be partly transferred from the shoulders of parents of large families to childless adults.[78]

This last recommendation was, among other things, a way of meeting a demand which had long been urged by eugenists: that bachelors should be punished for neglecting their racial responsibilities. Professor Edgar, of St Andrews University, told the International Eugenics Congress in 1912 that 'something should be done to encourage or compel the bachelors to undertake some of the responsibility of life. The shirking was exceedingly selfish, and it made things harder socially for the married'.[79] But Edgar's remedy, a tax on bachelors, might, Darwin feared, have racially undesirable consequences, since he believed that those 'who now remain unmarried are likely on the average to be somewhat inferior to the married'.[80] On the other hand, calculating income-tax liabilities on a 'family basis' would be a good way of stimulating the fertility of the sound stocks in the community, and would have no accompanying disadvantages.

But Darwin failed to convince the Royal Commission on the Income Tax that his main proposal was either desirable or workable.[81] Some of his suggestions were, indeed, adopted, in whole or in part, but this had little to do with the memorandum he submitted, and still less to do with eugenic considerations. The debate about child rebates had developed into a confrontation between differing conceptions of social justice. In this situation the ideas emanating from the E.E.S. were bound to appear, however strenuously Major Darwin sought to combat this impression, as yet another piece of special pleading from one more middle-class defence organization.

But by the 1920s, Darwin, along with most other eugenists, was less concerned with measures of positive eugenics than with restraining the multiplication of the unfit. What in practice did this demand involve?

8

Negative Eugenics

The most drastic solution would have been to kill off the 'unfit', however they may have been defined. Dr Chapple, writing for a New Zealand audience, quoted an American doctor who wanted idiots, imbeciles, epileptics, habitual drunkards, insane criminals, most murderers, nocturnal house-breakers, and other 'incorrigible' anti-social personages to be given a medical examination to see whether or not they should be exterminated: 'The painless extinction of these lives would present no practical difficulty—in carbonic acid gas we have an agent which would instantly fulfill the need'.[1] Such demands as these, sometimes heard in America, were never seriously put forward by British eugenists. An instructive episode occurred in November 1909, when the Mayor of Plymouth, on his own initiative, publicly advocated that, if three doctors decided that the hopelessly unfit and feeble-minded stood no possible chance of recovery, they should be painlessly put to death. Celebrities from various walks of life at once rushed into print, dissociating themselves from so wicked a proposition, and among them was the Chairman of the E.E.S., the unfortunately named Dr Slaughter.[2] Not the least of Shaw's offences in his notorious speech to the Society in March 1910 was his advocacy of 'Murder by the State', as the newspaper headlines put it: 'A part of eugenic politics', Shaw is quoted as saying, 'would finally land us in an extensive use of the lethal chamber. A great many people would have to be put out of existence simply because it wastes other people's time to look after them'.[3] Shaw probably did not expect his audience to take everything he said literally; yet this was widely felt to be a joke in the worst possible taste, and orthodox eugenists, not the quickest of people to appreciate even a good joke, were emphatically *not* amused.

Only slightly less drastic was the suggestion that defectives of one kind or another should be compulsorily sterilized. The most persistent advocate of this was the Liverpool physician, Robert Reid Rentoul, who published his *Proposed Sterilization of Certain Mental and Physical Degenerates* in 1903, and continued to reiterate his views over the next decade with a violence of language and a disregard of all the difficulties, which cannot have done his cause any good.[4] Other medical experts were on record as advocating sterilization as a means of stamping out gross physical deformities. Dr Yearsley, somewhat apologetically, suggested this as appropriate for congenital deaf-mutes, arguing that it would reduce the cost of maintaining these incurable cases in public institutions, so allowing more time and money to be spent on the 'acquired' cases.[5] Hereditary blindness, epilepsy, insanity, and imbecility were also occasionally discussed in this context.

But even Arnold White, who boasted of having coined the phrase, 'the sterilization of the unfit',[6] did not really believe it to be either practical or desirable to use surgery on those whom he categorised as 'degenerates'. 'England must choose between State suicide and race-improvement', he proclaimed, but on closer examination White is seen to be proposing relatively mild measures, like the permanent segregation of certain classes of 'defective', a revision of the marriage laws and even changes in the *social* environment.[7] If White shrank from advocating sterilization *tout court*, it may be readily imagined that the respectable professional men who made up the leadership of the Eugenics Movement moved with even greater care. Before 1914, the 'official' line was that put over by Professor Lindsay: 'It is to be hoped that sterilisation of the "unfit", which is at present being practised, with dubious results, in several States of the American Union, will not be pressed. Even if it could be justified, which is doubtful, public opinion in this country is not ripe for so drastic a proceeding. The public conscience would be shocked by it, and a promising movement would probably receive a rude check. Many feel instinctively that we might purchase a biological benefit too dearly at the cost of a spiritual wound'.[8] At the first undergraduate meeting of the Cambridge University Eugenics Society, Mr C. S. Stock, its Secretary, said, on the subject of sterilization: 'Well! we intend to follow the example of the London Society which regards any general discussion of this subject at present as futile and likely to be very misleading for the sufficient reason that the matter has not yet been exhaustively considered by a committee of Experts and pending such investigation we are more or less in the dark as to the physical, mental and moral effects of such action, not to mention the serious ethical problems that are raised . . .'[9]

Lindsay's reference to America is interesting. Here the pioneering work had been accomplished by Dr Harry Sharp who, from 1899 onwards, was performing vasectomy operations on his charges in the Indiana Reformatory, originally to prevent them from excessive masturbation; later Sharp convinced himself that this was a good way of preventing the procreation of mental defectives and of other sufferers from hereditary diseases. He had already operated on 223 inmates before the Indiana State Legislature legalized the proceedings in 1907. From then onwards it was mandatory for the authorities in this State to sterilize confirmed criminals, idiots, imbeciles and rapists, in public institutions, when recommended to do so by a board of experts.[10] By 1914, the example of Indiana had been followed by fifteen other American States, including California, where over the next two decades most of the sterilization operations were actually performed.

But even the American eugenists were divided over the desirability of this development, and, according to Haller, probably over a half of those sympathetic to eugenics in the 1910s and 1920s opposed the campaign for sterilization. Many of the State Bills were loosely drafted, and aroused antagonism within the legal profession by not making provision for notification of close relatives or for appeals to the courts, while doctors were unhappy at the haphazard collection of offences and medical conditions which made a citizen liable to be sterilized. In some States the legislation was openly punitive, and not based upon eugenic grounds at all. After the War, most American sterilization laws were revised, and the more glaring anomalies and injustices removed.[11] But already eugenists in Britain had been frightened off by what was happening on the other side of the Atlantic. As late as 1938, J. B. S. Haldane could dub compulsory sterilization 'as a piece of crude Americanism like the complete prohibition of alcoholic beverages'.[12]

Voluntary sterilization was, of course, a somewhat different matter. Havelock Ellis, who disliked the American sterilization laws, was attracted by the system in force in certain Swiss cantons of allowing defective persons in public institutions their liberty, in return for undergoing a sterilization operation; but the operation had to be performed with the consent of the subjects and of their relatives. This seemed to Ellis an experiment worth watching. The method to be adopted, he thought, was vasectomy for men and the ligature of the Fallopian tubes for women. But he was also excited at the prospect of sterilizing by means of X-rays, since this would not involve an operation at all, although the effects would last for several years.[13]

Many eugenists thought there was nothing wrong in offering the option of sterilization to people suffering from certain hereditary defects; one eugenic utopia of the time contains a touching picture of men thus mutilated being treated like soldiers disabled in the service of their country![14] To libertarians like Ellis, voluntary sterilization had the advantage that it would allow certain inmates of institutions to be let free. Some, but not all, asylum officers were impressed by the force of this argument.[15] Even Saleeby, who preferred to rely 'upon the creation of a eugenic conscience, and upon the self-control which we might hope [unfit] individuals would exercise', grudgingly admitted that there was a case for voluntarily sterilizing 'persons somatically normal but liable to transmit a genetic defect'.[16] In the 1920s, Major Darwin also pointed out that here was a cheaper method of preventing the multiplication of the unfit than placing them under custodial care.[17] This seems to have been one of the arguments which made converts to the cause of sterilization in the inter-war years, along with the recognition that the physical dangers of performing vasectomy operations were not high.

By the 1930s the Society was undertaking an enthusiastic campaign for voluntary sterilization. But before 1914 counsels of caution prevailed. For a start, it was pointed out that sterilization, like capital punishment, was an irreversible act, and an appalling situation would arise if the individual concerned or the appropriate authorities had a later change of mind. Secondly, it was feared that, whatever the authors of the proposal might intend, sterilization would turn out to be a 'class' measure, employed almost exclusively on the poor and on inmates of institutions.[18] Havelock Ellis had stressed that sterilization must be truly voluntary; but even he, despite his eloquent language about the dangers of compulsion, had argued that 'defective paupers of the second, third, or later generations' might perhaps be refused Poor Law relief, if they did not consent to undergo the preventative surgical treatment he favoured.[19] What sort of freedom was this? Finally, sceptics could point to alternative means of effecting the same end, which would avoid the dangers inseparable from sterilization and the risk of alienating public opinion.[20]

One of these possibilities was to prohibit marriage to men and women judged to be 'unfit'. Thus, Mrs Alec Tweedie urged that 'no-one should be allowed to marry without a doctor's certificate. It should be as necessary as the marriage lines'.[21] More subtle was Schiller's view that medical certificates should be voluntary, serving as a record of the above-average qualities, if any, possessed by their owners, but that the State should still have the right to refuse marriage to the grossly unfit.[22] A lengthy, unsigned article in the *Eugenics Review* for November 1910 advocated that 'those who perform the marriage ceremony, religious or civil, for the type of persons *found in the pauper population* should have power to refuse this on the basis of enquiry and full understanding of the facts of family history'. It might also be desirable, the article continued, to make some 'plan of insurance' an obligatory accompaniment of marriage.[23] This latter proposal merely repeated an earlier idea of Arnold White's, contained in *The Problems of a Great City* (1886): 'It is not too late to place legislative obstacles in the way of unions repugnant to a true sense of purity, hostile to national interests, and fraught with evil to the living and to generations unborn, by demanding from male minors evidence of means, before undertaking the burden of family life . . .'[24] In this and all his subsequent works White advocated, in addition, marriage certificates, the raising of the age of marriage to prevent 'child marriages', and a change of attitude from the clergy. The Churches also came in for some hectoring language from Dr Slaughter: 'If the Church is to grasp its modern opportunity, failing which there is little need of it, clearly it must utilise its two Sacraments of Confirmation and Marriage for their true purpose, namely in the interest of an idealism which recognizes the responsibility laid upon the present by the future'.[25] More temperately, Crackanthorpe suggested that the marriage banns could usefully be given a eugenic connotation, if the law stipulated that 'each of the contracting parties—the man on his own account, the woman by her parents or guardians—[were obliged] to make a solemn declaration' that there was no transmissible disease in the family.[26]

Ultimately, however, the responsibility for prohibiting marriages deemed likely to be 'dysgenic' would have to lie, not with the Church, but with the State, and this many eugenists felt to be unwise. The Whethams, for example, denied the right of the State to interfere with the freedom of choice of husbands and wives,[27] except in the case, presumably, of those who for one reason or another needed to be taken into custodial care. Secondly, any system of medical certificates would have imposed upon the medical profession an onerous and delicate responsibility that it had no wish to assume, as Squire Sprigge, editor of *The Lancet,* explained in a carefully worded article in the November 1909 number of the *Contemporary Review.*[28] Finally, there was the practical difficulty that briefly worried Dr Tredgold, who pointed out that illicit intercourse could not be prevented, and that the requirement of a certificate of health, whilst diminishing legal marriages, would almost certainly result in an increased illegitimate birth-rate.[29] This risk, however, was one which many eugenists thought was worth incurring; illicit intercourse between 'degenerates', they said, would be likely to occur, whatever the formal marriage conditions stipulated by the State. (Few eugenists went along with Dr Rentoul's extraordinary suggestion that extra-marital sexual intercourse should be made a criminal offence!)[30] However, *dissuading* the unfit from marrying was, of course, quite a different matter.

The main problem which eugenists set themselves, therefore, was that of altering the attitudes of the public towards marriage. The editor of the *Eugenics Review* declared: 'We hold that the essential method of eugenics is the ennobling of marriage, and thus regard any contrary tendency as undesirable in the highest degree'.[31] This ruled out of account the idea of the arranged marriage; instead, eugenists were quite prepared to allow sexual selection to be determined by love, instinct, the 'voice of nature', call it what you will, since they were in general agreement with Shaw's view that the existence of physical attraction between the sexes must have some biological function, and that 'falling in love' might perhaps be Nature's way of improving the human race.[32]

Yet at the same time eugenists contended that love was highly dirigible. In practice it was easy to show that the magnetic pull of sexual attraction was counteracted to a considerable extent by a large number of taboos and conventions, which operated all the more powerfully in that people were not usually conscious of them. Galton told the story of a lady belonging to an established county family who had 'scandalised her own domestic circle two generations ago by falling in love with the undertaker at her father's funeral and insisting on marrying him'. Galton commented: 'Strange vagaries occur, but considerations of social position and of fortune, with frequent opportunities of intercourse, tell much more in the long run than sudden fancies that want roots . . .'[33] There was also a distinction to be made, as Galton argued, between the 'two stages . . . of slight inclination . . . and falling thoroughly in love . . . If a girl is taught to look upon a class of men as tabooed, whether owing to rank, creed, connections, or other causes, she does not regard them as possible husbands and turns her thoughts elsewhere. The proverbial "Mrs Grundy" has enormous influence in checking the marriages she considers indiscreet'.[34] The argument that 'human nature would never brook interference with the freedom of marriage' was hardly borne out by knowledge of the sexual mores of different societies and tribes, which, as Galton demonstrated, exhibited a bewildering variety of conventions and marriage regulations.[35]

Unfortunately, many of the social conventions which governed courtship and marriage in modern industrial societies had no eugenic value, while some were positively harmful. The impecunious British aristocrat who sought to repair his family fortunes by marrying a 'barren' heiress was one example given by Galton of the existing tendency to value money more than 'good stock'. 'Every ostentatious wedding', said Saleeby, 'every luxurious, wasteful, pathological honeymoon, every newspaper paragraph which chronicles such things, is a lesson to young people that life and love and the future are words and money the only reality'.[36] There was something wrong, thought Schiller, with a society in which any man would rather marry a princess rather than a kitchen maid, though the latter might be prettier, healthier, and 'eugenically more commendable'.[37] This line of argument opened up the path for radicals like Shaw to advocate the panacea of an egalitarian society in which the entire community would be intermarriageable.[38] Orthodox eugenists, however, contented themselves with attacking the false values of society, in the hope that their own propaganda efforts would cause young people to re-think their attitudes towards matrimony, and to take a more clear-headed view of its biological purposes.

This involved, among other things, persuading young people to refrain from marriage if they came from families where there was a record of hereditarily transmissible disease. 'Genetic counseling', as we would now call it, probably goes back in Britain to the Edwardian decade. Dr Clouston, the pathologist, claimed that people were now disposed 'not only to asked medical advice about marriage and procreation but they actually follow it'.[39] But, unfortunately, doctors could not agree among themselves about what diseases were hereditary in their transmission. One physician, Harry Campbell, advised the following not to marry, even if they had recovered from the disease in question: sufferers from phthisis, organic heart disease, epilepsy, insanity, chronic Bright's disease, and perhaps rheumatic fever.[40]

But this was entering contentious ground, as is apparent to anyone who has followed the discussions that took place at a conference held in late 1908 under the auspices of the Royal Society of Medicine.[41] Was alcoholism hereditary? And what stand should be taken on diseases like tuberculosis and the zymotic diseases, which were not strictly speaking inherited, but to which certain stocks were much more resistant than others? Eugenists and doctors alike argued bitterly and inconclusively about such dilemmas as these. It is, perhaps, significant in this connexion that the E.E.S. Minutes for 1912 record that several letters had been received from members of the Society about medical examinations for marriage, but that the Council felt 'that any recommendation of individual medical men to give definite advice in any way connected with the Society as to the fitness of marriage of individuals would be extremely dangerous'.[42] In fact, the issue of marriage certificates threatened to sour relations between the eugenics movement and the medical profession. Doctors often expressed amused contempt at some of the uninformed eugenical prognostications to which they were subjected.[43] But eugenists, for their part, had good grounds for deploring the conservatism and 'amateurishness' of the medical profession; as J. B. S. Haldane has pointed out, it was not until as late as 1938 that genetics became a recognized part of the medical syllabus.[44] Nor did an 'entente cordiale' develop in these years between medicine and mathematics, as Karl Pearson had hoped.[45]

One final point needs to be made about a voluntary system of marriage control. As Darwin observed, the trouble with this system was that only those citizens who were alive to their racial responsibilities would go to the trouble of seeking medical advice before marrying, and would accept the expert advice that they were given; yet such people were *ipso facto* likely to be of some 'civic worth', even if they did have the misfortune to suffer from certain physical or mental ailments of a hereditary character.[46] The *real* degenerates, on the other hand, those whose rapid multiplication the eugenists were most anxious to prevent, would be in no way inhibited by anything less than legal sanctions, since they were, almost by definition, incapable of forethought, indifferent to their responsibilities to others, and immune to the pressure of informed and educated public opinion. The 'unfit', in other words, were unlikely to be subscribers to the *Eugenics Review*; how were these people to be reached?

Before answering this question, one obvious issue must be disposed of. So far it has been assumed that the unfit could be prevented from procreating by being prevented or dissuaded from marriage. But child-bearing and marriage can be separated. Voluntary sterilization is one way of doing this, although in Britain it would have necessitated legislation, since a doctor who performed a eugenic sterilization operation, even with the consent of the person operated on, laid himself open to prosecution on the grounds of mutilation.[47] But *contraception* was a method of controlling human reproduction well known to the Edwardian generation—so well known, in fact, that many eugenists believed that contraception had been responsible for the decline of the birth-rate in the first place. Even Pearson, who thought that the main reason was economic, added that the 'increased burden of parenthood for the mass of the population would not have led at once to its full consequences had not the trial of Mr Bradlaugh and Mrs Besant in 1877 resulted in a wide spread knowledge of the possibility of differentiating marriage and parentage'.[48] In truth, the supporting evidence for such assertions is thin. The Commission on the Declining Birth-Rate confirmed the findings made eight years previously by the Fabian Society, namely, that a substantial proportion of married couples from the middle classes, perhaps between forty-five and sixty per cent, were deliberately controlling the size of their families; but the Commission also reported that of those people in their survey (college graduates) who had given details of the methods employed, '51·7 per cent did not make use of any chemical or mechanical contrivance, but appear merely to have restricted marital intercourse to periods when conception was generally believed to be unlikely to occur or to have abstained altogether'.[49] Later investigators have concluded that, taking the community as a whole, the importance of contraceptives in pre-war Britain can easily be exaggerated, and that it was not until the inter-war years that their use became common.[50]

Nevertheless, contraception *could* have been employed as a deliberate eugenic device by those anxious about the rapid multiplication of the unfit. However, the issue of contraception was one which produced deep divisions within the eugenics movement. In general, those, who, for whatever reason, welcomed the advent of the small family also welcomed contraception, but even here there were exceptions; Dean Inge, for example, who expressed pleasure at the decline in the birth-rate, hastened to point out: 'I have no connection with the Malthusian League, I do not like their methods, and should never recommend them'.[51] Again, it was mainly 'reform eugenists' who came out in support of this new way of exercising a rational control over population growth. But amongst the leaders of the eugenics movement the highly conservative Montague Crackanthorpe had advocated what he called the 'Voluntary Principle' as far back as 1872, although he seems to have been referring to the 'rhythm' method of birth control.[52] Even more surprisingly we find Arnold White in

Problems of a Great City trying to rescue Neo-Malthusianism from its Radical and atheistic promoters: 'Limitation of families is a subject that has been spoiled by Mr Bradlaugh and his colleagues. In the public mind there is an indissoluble alliance between deliberate restriction and aggressive atheism. There is not, it is true, on the surface any necessary affiance between the two. Possibly Mr Bradlaugh and Mrs Besant are people who have lived a generation too soon. The arguments against them were employed against Sir J. Y. Simpson, when chloroform was first administered in cases of childbirth'.[53] It was a startling situation when Havelock Ellis could find himself for once in unity with a man like Arnold White, and yet on the opposite side of the fence from, for example, Dr Newsholme, the reforming Medical Officer of Health, who feared that artificial limitation of families could not 'be pursued on a large scale without moral loss to the community'.[54]

In general, the Eugenics Laboratory team took the sternest view about contraceptive appliances. Miss Elderton, who carried out a survey of family limitation north of the Humber, wrote in horrified tones of what she had discovered, and not only recommended a banning of the sale of abortofacients, but cried out for a 'real statesman' to come forth and warn the nation of the catastrophe that threatened to engulf it, if it did not desist from use of the 'contra-conceptive devices', which were so easily obtainable in the North of England.[55]

But few active members of the Eugenics Society adopted this alarmist position. In fact, as early as February 1909 those indefatigable Neo-Malthusians, Dr and Mrs Drysdale, were admitted into the Society, followed in 1912 by Miss Marie Stopes.[56] True, the Society Council declined Drysdale's suggestion that they send a delegate to the Neo-Malthusian Conference due to be held at The Hague in August 1910.[57] But this was no more than prudence. Sir James Barr, in a public lecture on eugenics, attempted to side-track a very contentious subject when he announced that he did not propose to touch the teaching of the Neo-Malthusians, which was 'more social and environmental than racial',[58] an assertion that was not altogether true, since Dr Drysdale, for one, does seem to have taken very seriously the possibility of using birth-control as an eugenic instrument.[59] At least Barr did not explicitly censure contraception, as many of his fellow doctors were doing, and this was to have important consequences; according to Dr Peel, the Malthusian League finally decided in 1913 to enter the field of medical propaganda, 'encouraged by the fact that Sir James Barr in his presidential address to the British Medical Association in the previous year, had advocated a low birth rate on eugenic grounds and had omitted to condemn contraception!'[60]

But in the pre-war years the E.E.S. could scarcely have undertaken this missionary work itself, even if its members had been virtually unanimous on the subject, which was far from being the case. The Churches were all officially hostile to the use of mechanical or chemical means of prevention, although the Church of England accepted, in 1914, the 'rhythm' method of family planning. Moreover, as late as 1909 the British Medical Association was sponsoring a Bill for making the sale of contraceptives illegal, and, in Dr Peel's words, doctors commonly alluded to their harmful effects, which allegedly included 'galloping cancer, sterility and nymphomania in women, and mental decay, amnesia and cardiac palpitations in men, [while] in both sexes the practice was likely to produce mania leading to suicide'.[61] It is rare, indeed, to encounter such hysterical nonsense in eugenical literature. The objection to contraception, where it existed, almost invariably rested on the ground that it was the better educated and the better informed who were using these appliances, while the feckless poor still carried on with their traditional habits: hence, the differential birth-rate. But this problem (assuming for the moment that it was a real problem) could theoretically have been tackled by eugenists publicizing contraceptives and conducting a propaganda campaign among the poor and the socially dependent. That, in fact, happened to a considerable extent after 1918, when in Britain, as in America, something of a rapprochement took place between the birth control movement and the eugenics movement.[62] Indeed, by the 1920s and '30s, the wheel had turned full circle. Instead of denouncing contraception as a filthy practice, harmful both to health and the future of the race, some eugenists, for example, Dr Cattell, were effectively arguing for *compulsory* contraception; that is, making public assistance dependent upon the recipients practicing birth-control, with friendly advice to be forthcoming from the relevant authorities, if necessary.[63] But before 1914 eugenists were understandably afraid of incurring additional ridicule and hostility, and it was left to Dr Drysdale, the Neo-Malthusian, to argue that 'if the poor had adequate opportunities of avoiding parentage, defective germ-plasm would be rapidly eliminated'.[64]

The easiest way out in the Edwardian period was thus for eugenists to urge that the unfit should be put into custodial care. Segregated from the rest of the community in special institutions, where the two sexes would be kept carefully apart, the unfit would be as effectively sterilized as if they had been operated upon. This expedient was obviously open to criticism for being expensive and for involving a curtailment of individual liberty, but it was likely to arouse less antagonism than sterilization proper or advocacy of contraception. Allies would also be forthcoming from outside the ranks of the eugenics movement, since the case for custodial care could also be argued in social and humanitarian terms, as well as in terms of the future of the race.

Segregation seemed to most eugenists an especially good way of dealing with the 'hereditary criminal', if indeed such a being existed. Thus, Galton had lent his authority to proposals for the 'prolonged separation of habitual criminals',[65] so as to limit their opportunities for producing 'low class offspring'. Prior to the publication of Goring's study, it had been commonly supposed that persistent offenders had especially large families. Dr Chapple thought that the unfit generally had defective inhibitions, which led them to commit crimes and at the same time made them prolific parents.[66] This hunch was not borne out by the statistics, which showed that among habitual criminals fertility was low, not just because of the long prison sentences which they served, but also because of the numerous cases of criminals being deserted by their wives.[67] Nevertheless, the fear that 'criminality' was a disease that was transmitted genetically from one generation to another enjoyed a long lease of life. And this explains why eugenists were attracted by the idea of using preventive detention as one method of sterilizing the unfit. A wide range of eugenists, from Arnold White to Sidney Herbert, favoured action along these lines.[68] What most of them wanted was for preventive detention, at present confined to *serious* offenders, to be applied also to *persistent* offenders, who, from the eugenic point of view, allegedly constituted the greater racial menace.[69] But, as we have seen, by 1911 or so it was widely accepted that a high proportion of 'habitual criminals' behaved in the way that they did because they were mentally retarded. And this brings us back to the important campaign for placing the feeble-minded in custodial care, a campaign in which eugenists played an influential part.

9

The Mental Deficiency Act, 1913

The impetus to reform came with the setting up of the Royal Commission into the Care and Control of the Feeble-Minded in 1904. The Government set this investigation on foot partly at the instigation of the prison and poor law authorities, who were worried at the cost of maintaining large numbers of defective persons whose failure to meet the demands of society seemed to be the consequence of arrested mental development.[1] But the scare about racial deterioration also played its part.

When the Royal Commission reported in 1908, Crackanthorpe wrote off to *The Times*, claiming that the principles of eugenics had 'found their way into a Blue-book, [and] eugenics may be said to have become "nationalized" by authority. Certainly, no social reformer, to whatever political party he belongs, or belonging to none at all, can afford to neglect its earnest study'.[2] Since one of the medical experts upon whom the Royal Commission had relied was Dr Tredgold, soon to become a leading member of the E.E.S., it is not suprising that its recommendations were to Crackanthorpe's liking. The Commissioners made it clear that they viewed feeble-mindedness as a hereditary condition, noted that the feeble-minded, as a class, had a fertility well above the average, and proposed that certain categories of the feeble-minded would have to be segregated from the rest of the community, in the interests of others as well as for their own protection.[3] This proposal was endorsed in early 1909 by both the Majority and Minority Poor Law Commissioners, and eugenists were sanguine that the government would soon introduce the appropriate legislation.

9 *The Mental Deficiency Act, 1913*

Questions were being asked in the Commons about the government's intentions in February 1909. But Herbert Gladstone's reply was ominous; the government's actions, he said, 'must be subject to Parliamentary exigencies'.[4] The Constitutional Crisis, which broke in earnest later that year, was bound to hold up a settlement of the feeble-minded question. In June 1910, Churchill, the new Home Secretary, assured the House that a draft Bill was already in preparation, but that further legislation that Session was impractical.[5] The Commons elected in December 1910 soon became absorbed in the fate of the Parliament Bill, the National Insurance Bill, and, later, the re-opening of the Home Rule dispute. As early as 1910, Crackanthorpe was complaining that, despite the recommendations of the recent Royal Commission, the 'much needed legislation . . . appear[ed] to be postponed to the Greek calends', 'owing to that fetish, Political Party, which leads so many good men astray'.[6] It is, of course, hardly surprising that the Liberal Administration should have accorded low priority to an issue which did not figure in the Party programme and was unlikely to be popular with the party activists. Indeed, even at this early date, there were rumblings of protest from a group of doctrinaire individualists on the government backbenches, led by Josiah Wedgwood, alarmed at what they saw as a new threat to the 'liberty of the subject'.

But Churchill himself, the Home Secretary between February 1910 and October 1911, cannot be accused of not taking a sympathetic interest in the matter. In December 1910, he wrote a letter drawing the Prime Minister's attention to the 'multiplication of the unfit', which constituted, he said, 'a very terrible danger to the race'. He informed Asquith that until the public came round to accepting the need for sterilizing operations, the feeble-minded would have to be kept in custodial care, segregated from the world and from the opposite sex.[7] Sometime in 1910 Churchill also circulated among his Cabinet colleagues an address which Tredgold had delivered the previous year, entitled 'The Feeble-Minded—A Social Danger'. In a covering note Churchill endorsed Tredgold's alarmist presentation of the case: 'The address gives a concise, and, I am afraid, not exaggerated statement of the serious problem to be faced. The Government is pledged to legislation, and a Bill is being drafted to carry out the recommendations of the Royal Commission'.[8] When a deputation petitioned the government in October 1910 to act without delay, it may be significant that, whereas Asquith and Loreburn were cautious and non-committal, Churchill, in reply, 'recalled the fact that there were at least 120,000 feeble-minded persons at large in our midst who deserved "all that could be done for them by a Christian and scientific civilization now that they were in the world", but who should, if possible, be "segregated under proper conditions so that their curse died with them and was not transmitted to future generations"'.[9] This was strong language, and went further than was necessary or even prudent from the government viewpoint, and it lends some credibility to the following entry in W. S. Blunt's Diaries, dated October 1912:

> Winston is also a strong eugenist. He told us he had himself drafted the Bill which is to give power of shutting up people of weak intellect and so prevent their breeding. He thought it might be arranged to sterilize them. It was possible by the use of Röntgen rays, both for men and women, though for women some operation might also be necessary. He thought that if shut up with no prospect of release without it many would ask to be sterilized as a condition of having their liberty restored. He went on to say that the mentally deficient were as much more prolific than those normally constituted as eight to five. Without something of the sort the race must decay. It was rapidly decaying, but would be stopped by some such means.[10]

But by October 1912 Churchill had moved to the Admiralty, and his successor at the Home Office, Reginald McKenna, had still not produced the legislation upon which the Department had allegedly been working since 1910. Pressure would have to be applied on the government.

The two bodies to organize the campaign were the National Association for the After-Care of the Feeble-Minded and the Eugenics Education Society itself. The former body was more concerned with protecting the feeble-minded, as individuals, the eugenists with the danger of 'racial decay', but the work of the two organizations overlapped, and many of those prominent in the agitation belonged to them both. In October 1910, the National Association and the E.E.S. joined forces in sending a deputation to see the Prime Minister.[11] In March 1911, the National Association went further, by proposing co-operation with the E.E.S. in the preparation of a short Bill, that could be introduced by a private member; the Bill should 'be short and simple, granting power of detention to existing, duly recognised institutions for inmates of 16 years and over'.[12] The Council of the E.E.S. took up the suggestion with alacrity, and a Bill was quickly drafted. In May, the National Association were proposing, on the strength of an assurance from the government that the Home Office had the matter well in hand, to drop their own Bill, but their resolution was stiffened by the Council of the E.E.S., which persuaded them that the measure should be brought forward, regardless.[13]

The next stage in the campaign was the convening by the two bodies, in December 1911, of a meeting at the House of Commons to explain the contents of their Feeble-Minded Persons (Control) Bill to M.P.s of all parties. This meeting was arranged by the Liberal Member, Walter Rea, one of the few M.P.s to take out membership of the E.E.S. Its outcome was the formation of a Committee of M.P.s to watch over the interests of the Bill.[14] In March 1912 Rea was able to report that this Committee had reached the decision that a Unionist backbencher, Gershom Stewart, should adopt the Feeble-Minded Control Bill and introduce it into the Commons on 17 May. At once an interview was arranged between Stewart and Major Darwin.[15] The briefing session seems to have been effective, since Stewart's speech of introduction was a concise rehearsal of arguments and factual evidence very familiar to anyone who has perused the eugenical literature of the day.[16]

The Private Member's Bill passed its second reading unopposed, and, although it was clearly not going to make further progress, it had served its purpose of flushing the government out into the open. McKenna intervened in the second reading debate to announce that the Home Office Bill was now ready to come before Parliament.[17] All this while a lively campaign, organized by the National Association and the E.E.S., was being conducted in the country, to keep the government up to the mark. In December 1912, Ellis Griffith, Under-Secretary at the Home Office, admitted that his Department had received about eight hundred resolutions in favour of legislation.[18] There can have been few towns of any size which did not hold a public meeting in the 1909–13 period, to protest at the Government's passivity in the face of the 'menace of the feeble-minded'.[19]

The government's Bill, the Mental Deficiency Bill, was at last published, and passed its second reading on 19 July 1912 by the margin of 230 votes to 38. But it was then strangled in standing committee; persistent hostility from a small but determinded group of backbenchers, unsatisfactory drafting which bothered even M.P.s who supported the principle of the Bill, and lack of Parliamentary time all led to the dropping of the measure at the end of the Session. There were more protest meetings, and on 28 November *The Times* published a letter, carrying a long list of distinguished signatories, which argued that the delay was involving cruelty to the feeble-minded themselves and danger to others, since the feeble-minded 'leave behind them a new generation of mentally and physically degenerate children, not only continuing, but increasing, the number who must be supported at the expense of the community'.[20]

In one sense the 1912 Bill had perhaps been too satisfactory from the eugenists' point of view, and had in consequence aroused opposition which reduced its chance of reaching the statute book. Wedgwood, the most persistent critic, was particularly aghast at Clause 17, which permitted feeble-minded persons to be placed in custodial care when 'it [was] desirable in the interests of the community that they should be deprived of the opportunity of procreating children'; Wedgwood called this 'the most abominable thing ever suggested', and hinted that the sponsors of the Bill wished to sterilize defectives.[21] A clause making it a misdemeanour to marry a defective was additional proof, to men like Wedgwood, that the whole agitation was tainted with the 'spirit of the horrible Eugenics Society which is setting out to breed up the working classes as though they were cattle'.[22] Outside Parliament, G. K. Chesterton and Hilaire Belloc, through his paper, *The Eye-Witness*, were making similar allegations.[23]

Thus, when McKenna introduced the Mental Deficiency Bill in a new form in 1913, he went out of his way to play down any connections it might have had with the E.E.S. Gone was the controversial section of the old Clause 17, and gone, too, the prohibition of marriage with a defective. Dr Chapple, a Liberal M.P. who, when a doctor in New Zealand, had written a hair-raising book about eugenics, *The Fertility of the Unfit*, did his best, but with only limited success, to get a definite eugenical content back into the Bill.[24] McKenna himself stoutly insisted that the measure, as it now stood, existed only for the protection of individual sufferers.[25] But Wedgwood, hyper-sensitive to the machinations of 'eugenics cranks', was hard to convince, and Chapple's activities made him doubly uneasy. It is noticeable, however, that by July 1913 even Chapple felt obliged to protest that he was actuated, not by a concern for the race, but solely by a concern to help the feeble-minded themselves.[26] Public espousal of eugenical doctrines was still politically hazardous for an M.P.

The Mental Deficiency Act passed its second reading in June 1913, with the only opposition coming from eleven refractory M.P.s—stern individualists from the extreme Radical Left, like Wedgwood and Outhwaite, and traditionalists from the right of the Unionist Party, like Banbury. The measure received the Royal Assent on 15 August, and was due to take effect on 1 April, 1914. Although the Bill had been toned down since its first introduction the previous year, the E.E.S. could feel considerable satisfaction. The Home Office had consulted them in the re-drafting of the Bill.[27] And, however modified, the compulsory powers to detain and segregate the feeble-minded had largely met the substance of the eugenists' demands for the curbing of the multiplication of the unfit. The Mental Deficiency Act, to quote the *Eugenics Review*, was 'the only piece of English social law extant, in which the influence of heredity has been treated as a practical factor in determining its provisions'.[28] In the spring of 1914, the E.E.S. was confidently predicting that the Elementary Education (Defective and Epileptic Children) Bill and an Inebriates Bill, then before Parliament, might shortly become law, to supplement what had already been achieved.[29] But the First World War seriously interrupted the progress of eugenics; and the E.E.S., its membership dispersed and its provincial branches disbanded, did not recover its former effectiveness until well into the 1920s. When in the early 1930s eugenics once again became politically significant, it did so within a context that was quite different from the 1908–14 period, and this phase of its history deserves separate treatment in another place.

10

Conclusion

Why should eugenics have proved so widely attractive in Britain on the eve of the First World War? It is tempting to argue that the rise of the eugenics movement reflected structural changes in the economy taking place in these years. Professor John Rex, a Marxist scholar, has concluded an article attacking the objectivity of I.Q. tests with the observation that 'the mis-representations of the psychometricians is not simply a matter of a random "mistake" but is directly related to the beliefs of the society in which they operate'.[1] The same line of reasoning could be applied to eugenics as a whole, which historically has depended to a considerable extent on psychometric research. Supporting evidence for such an interpretation is contained within this monograph. Many eugenists, as we have seen, presented their creed as a validation of class inequalities, hoping thereby to stabilise an economic order beginning to come under attack from Radical politicians and labour militants.

But quite apart from the difficulties of matching up the fortunes of the British eugenics movement with economic change, such an approach runs into several difficulties. For a start, leading eugenists earlier in the century were not, as a group, especially enamoured of contemporary forms of industrial capitalism. There were actually those, like Dean Inge, to whom industrialism was a cancerous growth on British society, and who eagerly looked to the eventual dismantling of the manufacturing cities so that their sites could be reclaimed for the plough.[2] Although this was an extreme position, it is noticeable that very little support for eugenics came from 'captains of industry', bankers, or businessmen.[3] Nor, conversely, were all socialists hostile.

Indeed, as contributors to the *New Age* were arguing in the Edwardian period, socialism and eugenics (stripped of its crass class prejudices) could well be regarded as complementary rather than competing programmes.[4] For those socialists prepared to base their case on the essential *inequality* of man, eugenics has proved to be extremely attractive, as can be seen from a study of the writings of socialist intellectuals from George Bernard Shaw to Professor J. B. S. Haldane, both of them 'reform eugenists' of a sort. True, trade union leaders have mostly ignored eugenics, and often seemed unaware of its very existence. It still remains a matter of some interest that the fiercest opposition to eugenics has come, not from the Labour/Socialist camp, but from Roman Catholics and from a certain kind of individualist liberal.[5]

Yet one ought not to push this line of argument too far. The main support for eugenics may have come less from the business community than from professional and academic circles, but it was fear of the growing power of the Labour Movement which directed the behaviour of a majority of the original adherents. Significantly, the pioneers of genetics in Britain, almost without exception, were from comfortable middle-class backgrounds, and class prejudice all too often crept into their evaluation of their scientific work.[6] But the same bias is to be found among the larger body of eugenists, the non-scientists included.

More specifically, the growth of the eugenics movement can be seen as one particular response to the emergence of social welfare politics. Eugenics proved to be a highly effective weapon for the belabouring of the pre-war Liberal Government. Agitated conservatives were anxious to find arguments against the Radical programme other than traditional appeals to individualism and self-help, which by this time had acquired a decidedly old-fashioned flavour, and eugenics admirably filled this need. Lloyd George and Churchill could now be attacked, not simply as Radical mischief-makers working to set class against class, but as ignorant men attempting to set the laws of biology at nought, or as 'bad stockmasters'. Also, eugenics supplied a much-needed 'constructive' alternative both to the Liberal programme and to socialism: and one that, so its supporters claimed, went to the root of the evil and did not deal solely with surface manifestations. Finally, eugenics was very useful to people of a conservative disposition, because it could be invoked in the defence of traditional economic theories; if these theories seemed not to be in accord with the known facts of social life, eugenists could explain this discrepancy as one of the consequences of 'physical deterioration'. First implement the eugenic programme for the improvement of the human stock, they suggested; the major 'impediment' to the proper working of economic laws would then have been removed.

Support for eugenics may also have come about, in part, in reaction against the facile 'environmentalism' of some late Victorian social reformers. Mark Haller, for one, seems to be implying that in America 'hereditarian' explanations of social problems, like crime, may have found favour because of disappointment that criminals were not 'responding', as sanguine reformers had expected they would, to the improved facilities now being provided by the public authorities.[7] Perhaps social workers, public officials, and their professional advisers were thus, however innocently, looking for arguments to discount their own relative failure. Such a hypothesis cannot, obviously, be proved true or false, and supporting evidence is probably not as abundant in Britain as it is in America. But we have already encountered several prominent eugenists who possibly fit Haller's description, such as E. J. Lidbetter (a Poor Law relieving officer) and Dr F. W. Mott (Pathologist to the London County Asylum).

It might also be suggested that eugenics was likely to make an especially telling appeal in a society where science was held in very great respect, but where few civil servants and still fewer politicians had received a scientific education. One certainly senses, in many eugenical pronouncements coming from scientists, an acute irritation with the common assumption that a humanist education provided the most appropriate preparation for a career of public service. Eugenics seemed to demonstrate beyond doubt that such an assumption was false. But it could simultaneously be used to indict a political system that allegedly conferred power on 'amateurs'. Since the early nineteenth century there has been a persistent hope that politics and public administration might be made a 'science' in which objectively correct decisions could be made, on the basis of information supplied by the relevant 'expert'. Eugenics is only one of the many forms through which this 'technocratic' approach to politics has found expression.

The changing fortunes of eugenics can also be explained, however, not simply as a response to external political stimuli, but in terms of intellectual developments within the academic world. Haller has asserted: 'If the strength of eugenics lay in the support of the experts, its decline came as experts abandoned it'.[8] The challenge to the 'psychometric' investigations of Cyril Burt and his school, which goes back to the 1920s, provides an example of what Haller had in mind. But in Britain it was perhaps less a case of scientists 'abandoning' eugenics than of their so modifying its contents that they destroyed its popular appeal, and its status as a plausible political programme. Thus the discovery by geneticists that the mechanism of inheritance among human beings was more complex than had once been claimed, may have slowed down the growth of the eugenics movement—and also, incidentally, created a problem of communication between its 'scientific' and 'lay' membership, which had not existed in the innocent years before the First World War.

Equally important has been increasing specialization, which has done much to destroy the synthesis that eugenics once embodied. The attractiveness of the subject earlier in the century lay in the way in which it fused a variety of intellectual and practical pursuits into a more or less coherent programme, which both offered a plausible explanation of man's *past* development and a practical policy for the *future*. This programme has now, despite the efforts of the Eugenics Society to counter the trend, largely broken down into a number of specialized sciences (like demography) and professional occupations (like genetic counselling), which have tended to develop along independent lines. Finally, since the Second World War there has been a very deep-seated prejudice against any attempt to formulate a social policy shaped by 'biological' considerations. This factor, plus the complete eclipse of Social Darwinism as a language of philosophical and political discourse, has put obstacles in the path of eugenists which the founders of the movement could never have foreseen.

Appendix

Biographical Details Concerning Certain Prominent Eugenists

Crackanthorpe, Montague Hughes (1832–1913). Barrister. Owner of about 6,000 acres in Westmoreland. Educated at Oxford University. Member of General Council of the Bar and of Council of Legal Education. Member of International Commission on Criminal Sentences. Author of many articles on legal, social, and political topics published in the monthlies and in the *Encyclopaedia Britannica*. President of E.E.S., 1909–11.

Chapple, William Allan (1864–1936). Surgeon and politician. Born in New Zealand, where he practised as a surgeon, 1899–1906. A member of the New Zealand Parliament. Settled in the United Kingdom in 1906. Liberal M.P. for Stirlingshire, 1910–18, and for Dumfriesshire, 1922–4. Author of several books on medicine, physical development and education.

Darwin, Major Leonard (1850–1943). Fourth son of Charles Darwin. Educated at R.M.A., Woolwich, and entered Royal Engineers in 1871. On Staff of Intelligence Department at War Office, 1885–90. Retired from Army in 1890. Served on several scientific expeditions. Liberal Unionist M.P. for Lichfield, 1892–5, and Liberal Unionist candidate for that constituency in 1895 and 1896. President of Royal Geographical Society, 1908–11, and of the E.E.S., 1911–28. Chairman of Bedford College, 1913–20. Author of *Bimetallism* (1898) and *Municipal Trade* (1903), as well as of books and articles expounding eugenics.

Inge, Very Rev. William Ralph (1860–1954). Anglican clergyman. Educated at Cambridge University. Fellow of King's College, Cambridge, 1886–7, and Fellow and Tutor of Hertford College,

Oxford, 1889–1904. Lady Margaret Professor of Divinity at Cambridge, 1907–11. Dean of St Paul's, 1911–34. Prolific lecturer and essayist on current affairs and popular religion.

Pearson, Karl (1857–1936). Mathematician. Educated at Cambridge University. Fellow of King's College, Cambridge, 1875–9. Called to Bar, 1882. Director of the Eugenics Laboratory, 1907–11, and subsequently Emeritus Professor of Eugenics at London University. Editor of *Biometrika*, 1902–35, and of *Annals of Eugenics*, 1925–33. Biographer of Francis Galton. Author of numerous books and articles, most of them concerned with the mathematical theory of evolution, and of the popular *Grammar of Science* (1899).

Saleeby, Caleb Williams (1878–1940). Medical journalist. Educated at Edinburgh University. Practised medicine. Prolific author and (after 1924) broadcaster on medical subjects. Royal Institution Lecturer on eugenics, 1907, 1908, 1914, 1917, 1923. Founder and Chairman of the Sunlight League, 1924. Member of National Temperance League and of the National Birthrate Commission (being Chairman of the latter body, 1918–20). Unpaid adviser to Lord Rhondda at the Ministry of Food, 1917–18.

Schiller, Ferdinand Canning Scott (1864–1937). Philosopher. Born of German father. Educated at Rugby and Balliol. Instructor in philosophy at Cornell University, 1893–7. Tutor in philosophy and Fellow of Corpus Christi College, Oxford, 1903–26. Professor of Philosophy, University of Southern California, 1929. In 1935 married an American wife and took up permanent residence in U.S.A. Author of numerous books and articles expounding his theory of philosophical pragmatism.

Slaughter, John Willis (1878–1964). Sociologist. See p. 16 and the appended note, Chapter 2, n. 32, p. 122.

Tredgold, Alfred Frank (1870–1952). Neurologist. Educated at Durham University. Conducted researches into mental defects, 1899–1901. Medical expert to Royal Commission on Feeble-Minded, 1905. Physician in neurology, Royal Surrey County Hospital. Fellow of Royal Society of Medicine. Member of Board of Education Mental Deficiency Committee and of the Ministry of Health Committee on Sterilisation (Brock Committee), 1932–3. Author of the frequently re-issued *Mental Deficiency* (1908).

Whetham, William Cecil Dampier (1867–1952). Physicist. Educated at Trinity College, Cambridge. Fellow of Trinity, 1891, and Lecturer in natural sciences, 1895–1922. F.R.S., 1901. His many scientific books include *The Foundations of Science* (1912). His books and articles on eugenical subjects were mostly written in conjunction with his wife, Catherine, *née* Durning. His later years were devoted to farming operations and the study of agricultural economics. Secretary of Agricultural Research Council, 1931–5. Development Commissioner, 1933–51. Member of Central Agricultural Wages Board, 1925–42.

White, Arnold (1848–1925). Journalist. Unsuccessfully contested a seat in Parliament as a Liberal in 1885, as a Liberal Unionist in 1892 and 1895, and as an Independent in 1906. One of the leaders of the agitation to restrict Jewish immigration into the U.K., 1885–1905. Active member of the Navy League, and friend of Admiral Sir John Fisher. From 1907 onwards wrote regular column in *The Referee* under pseudonym, 'Vanoc'; also contributed regularly to a wide range of newspapers and journals.

Notes and References

In all references the place of publication is London, unless it is stated otherwise. When short-title references are used, the full citation will be found earlier within the same chapter. The following abbreviations are used:

ER	*Eugenics Review*
EES	Eugenics Education Society
Galton, *Essays*	Francis Galton, *Essays in Eugenics* (London, 1909)
Pearson, *Life*	Karl Pearson, *The Life, Letters and Labours of Francis Galton*, 3 vols(Cambridge, 1914–30), I(1914), II(1924), IIIA & IIIB (1930).
Problems in Eugenics	*Problems in Eugenics. Papers Communicated to the First International Eugenics Congress held at The University of London, July 24th to 30th, 1912*, 2 vols (London, 1912–13).

Introduction

1. Francis Galton, 'Eugenics: Its Definition, Scope and Aims', *Nature*, 70 (1904), 82; also in Galton, *Essays*, p. 35.
2. Caleb Williams Saleeby, *Parenthood and Race Culture: An Outline of Eugenics* (1909), p. 182.
3. John Higham, *Strangers in the Land: Patterns of American Nativism, 1860–1925* (New York, 1963), pp. 150–1.
4. Mark H. Haller, *Eugenics: Hereditarian Attitudes in American Thought* (New Brunswick, 1963). See also Donald K. Pickens, *Eugenics and the Progressives* (Nashville, Tennessee, 1968).

Chapter 1

1. Sir James Barr, *The Aim and Scope of Eugenics* (Edinburgh, 1911), p. 15.
2. [Henry] Havelock Ellis, *The Problem of Race-Regeneration* (1911), p. 67.

3. *Illustrated London News*, 12 March 1910.
4. F. C. S. Schiller, *Eugenics and Politics* (1926), p. vii.
5. Charles Wicksteed Armstrong, *The Survival of the Unfittest* (1927), p. 169.
6. M. Dachert, the founder of a eugenic community which flourished in Strasbourg in the inter-war years, is said to have derived his ideas straight from *The Origin of Species*; see *ER*, 23 (1931–2), 3.
7. Cited in G. Himmelfarb, *Darwin and the Darwinian Revolution* (1959), p. 351.
8. One of Darwin's sons, the botanist, Francis Darwin, said in the course of his Galton Lecture in February 1914 that his father, in the first edition of *The Descent of Man*, 'distinctly gives his adherence to the eugenic idea by his assertion that man might by selection do something for the moral and physical qualities of the race'; see *ER*, 6 (1914–15), 16.
9. John Humphrey Noyes, 1811–86. See the entry in *Dictionary of American Biography*.
10. Mark H. Haller, *Eugenics: Hereditarian Attitudes in American Thought* (New Brunswick, 1963), pp. 37–8.
11. But, while quoting Galton's scientific work, Noyes attacks Galton in this book for not advancing a practical programme; when it comes to what should be done, Galton, he claims, 'subsides into the meekest conservatism'. See J. H. Noyes, *Essay on Scientific Propagation* (Oneida, New York, [1875?]), p. 8.
12. Havelock Ellis, *The Task of Social Hygiene* (1912), pp. 29–30.
13. Many of her 'scientific' theories, in fact, ran counter to the central propositions of the eugenists, e.g., 'The most active agent in generating the unfit is fatigue poison . . .'; much 'family degeneration . . . is due to physical exhaustion from overwork or the lack of sufficient light and fresh air'; see V. C. Woodhull-Martin, *The Rapid Multiplication of the Unfit* (1891), p. 10. It is Mrs Woodhull-Martin's oft-repeated contention that those who were unfit through fatigue produced degenerate *offspring*. Although she lived the latter part of her life in London, where she edited a weekly, *The Humanitarian*, H. G. Wells is the only British writer known to me to acknowledge any debt to her; see his *Mankind in the Making* (1903), p. 39.
14. F. Galton, *Memories of My Life* (1908), p. 300.
15. On Galton, see C. P. Blacker, *Eugenics: Galton and After* (1952), part I; Pearson, *Life*; D. W. Forrest, *Francis Galton: The Life and Work of a Victorian Genius* (1974).

Chapter 2

1. Galton, *Essays*, pp. 1–34.
2. ibid., pp. 34–43.
3. Edgar Schuster, *Eugenics* (1912), 46–7; D. W. Forrest, *Francis Galton: The Life and Work of a Victorian Genius* (1974), pp. 260, 269–70.
4. A brief account of the early days of the Society is to be found in Faith Schenk &

A. S. Parkes, 'The Activities of the Eugenics Society', *ER*, 60 (1968), 142–61.
5. *Birmingham Daily Post*, 16 November 1907.
6. EES Minutes, 9 December 1907 and 14 January 1908.
7. *Pall Mall Gazette*, 27 February 1908.
8. Schenk & Parkes, 'The Activities of the Eugenics Society', p. 143.
9. *EES Annual Report, 1908*, pp. 21–5.
10. ibid., *1913–14*.
11. *ER*, 5 (1913–14), 1–64.
12. *William Bateson, F.R.S. Naturalist; His Essays and Addresses, together with a Short Account of his Life by Beatrice Bateson* (1928).
13. Editorial in *British Medical Journal*, no vol. no. (23 August 1913), 508–9.
14. *EES Annual Report, 1911–12*, pp. 23–5.
15. On the relationship between the early stage in the development of British sociology and eugenics, see Philip Abrams, *The Origins of British Sociology: 1834–1914* (Chicago, Illinois, 1968), and R. J. Halliday, 'The Sociological Movement, the Sociological Society and the Genesis of Academic Sociology in Britain', *Sociological Review*, 16 (1968), 377–98.
16. *Nature*, 84 (1910), 431.
17. *ER*, 5 (1913–14), 385.
18. See Records of the Cambridge University Eugenics Society, housed in the Library of the Eugenics Society, 69 Eccleston Square, London SW1V 1 PJ. Keynes, in fact, delivered the Galton Lecture in 1937.
19. H. J. Laski, 'The Scope of Eugenics', *Westminster Review*, 174, (1910), 25–34; Pearson, *Life*, IIIB, 606, 608–9; [Basil] Kingsley Martin, *Harold Laski (1893–1950) A Biographical Memoir* (1953), pp. 8–12.
20. *EES Annual Report, 1912–13*, pp. 5–6.
21. The controversial address by Shaw referred to, was delivered at a meeting organized by the EES and held at Caxton Hall in March 1910; see report in the *Daily Express*, 4 March 1910. The Society's President, Montague Crackanthorpe, took the unusual step of dissociating himself from Shaw's pronouncement in his address to the next A.G.M.; see *ESS Annual Report, 1909–10*, p. 1.
22. A similar thing happened in the U.S.A.; see Mark H. Haller, *Eugenics: Hereditarian Attitudes in American Thought* (New Brunswick, 1963), pp. 85–6.
23. *Problems in Eugenics*, II, 70.
24. *EES Annual Report, 1912–13*, p. 6.
25. Sidney Webb, 'Eugenics and the Poor Law: The Minority Report', *ER*, 2 (1910–11), 233–41, especially p. 240.
26. L. Darwin, 'The Aims and Methods of Eugenical Societies', in *Eugenics, Genetics and the Family: Volume I, Scientific Papers of the Second International Congress of Eugenics held at American Museum of Natural History, New York, September 22–28, 1921* (Baltimore, Maryland, 1923), pp. 5–19.

27. Nor were Galton and Pearson reassured when they read in the newspapers that one of the founder members of the Council had been convicted of indecent assault on a bus conductor. The person in question, Dr Stanton Coit, was later acquitted on appeal to Sessions; see Pearson, *Life*, IIIA, 335–6; and *The Times*, 26 March 1908.
28. EES Minutes, 3 March 1909. See also Forrest, *Francis Galton*, p. 278.
29. W. Marshall, 'Montague Crackanthorpe: Obituary', *ER*, 5 (1913–14), 352–3.
30. ESS Minutes, 1 February 1911 and 1 March 1911.
31. ibid., 1 March 1911.
32. He later returned to America, then served as a visiting lecturer to the Central China University, before taking up a post of lecturer in civics and philosophy at Rice Institute, Houston, Texas. He died in Houston, on 14 November 1964; see *Who Was Who in America*.
33. *The Times*, 27 March 1943.
34. The scientific background to this controversy has been ably analysed in P. Froggart & N. C. Nevin, 'The "Law of Ancestral Heredity" and the Mendelian-Ancestrian Controversy in England, 1889–1906', *Journal of Medical Genetics*, 8 (1971), 1–36.
35. EES Minutes, 12 February 1908.
36. David Heron, for example, resigned from the Council of the Society in 1910; see EES Minutes, 2 March 1910.
37. *Nature*, 92 (1914), 606, 660–1.
38. Pearson, *Life*, IIIA, 405–9.
39. C. W. Saleeby, *The Progress of Eugenics* (1914), p. 20. Review of Archibald Reid's *Principles of Heredity* in *Daily Chronicle*, 4 June 1910.
40. Pearson, *Life*, IIIA, 405–7.
41. Francis Darwin, 'Francis Galton, 1822–1911', *ER*, 6 (1914–15), 1–17 (p. 13).
42. K. Pearson & E. Elderton, *The Relative Strength of Nurture and Nature*, second edition (1915), p. 32.
43. Haller, *Eugenics*, pp. 67–8.
44. David Heron, *Mendelism and the Problem of Mental Defect: I, A Criticism of Recent American Work*, Questions of the Day and of the Fray No. 7 (1913).
45. *ER*, 5 (1913–14), 367.
46. Galton read this book in proof. His comment to Pearson was: 'though there is much I would myself strike out, [I] expect it will do good. He has eminently the art of popular writing with fluency . . .'; see Pearson, *Life*, IIIB, 597. Galton's toleration of Saleeby, however, was to disappear almost entirely over the next twelve months.
47. For example, 'It is to Dr Saleeby's shoulders that Mr Galton's mantle has been transferred . . .'; *Glasgow Herald*, 22 May 1909.
48. Galton wrote to Pearson on 7 May 1910 as follows: 'The Council of the Eugenics Education Society have, I learn, extruded Dr S. by not putting his name on the candidate list. As I am told, certain members of the Council

strongly objected to serving longer with him, and Mrs G [otto] undertook to tell him so, which she did, doubtless with all practicable tact, but I have reason to know his feelings are much wounded'; see Pearson, *Life*, IIIA, 428. But the Society's Minutes for 5 May 1910 indicate that Saleeby's name went forward for election, but that he was one of the unsuccessful candidates.

49. EES Minutes, 15 October 1913. Saleeby caused further embarrassment by then indicating his willingness to lecture the Society on any subject it chose; ibid., 21 November 1913.
50. Saleeby, *Progress of Eugenics*, passim.

Chapter 3

1. G. R. Searle, *The Quest for National Efficiency, 1899–1914* (Oxford, 1971), passim.
2. Arthur Newsholme 'Alleged Physical Degeneration in Towns', *Public Health*, 17 (1905), 292.
3. Gareth Stedman Jones, *Outcast London* (Oxford, 1971), especially chapter 6.
4. This point has been made by Philip Abrams, *The Origins of British Sociology: 1834–1914* (Chicago, Illinois, 1968), p. 124.
5. *Hansard*, 4th series, 160, cols 1392–5, 16 July 1906.
6. For example, W. C. D. Whetham & C. D. Whetham, *An Introduction to Eugenics* (Cambridge, 1912), p. 37; L. Darwin, *The Need for Eugenic Reform* (1926), p. 62.
7. *The Times*, 18 June 1909.
8. Quoted from Local Government Board report in *ER*, 1 (1909–10), 143.
9. According to Dr Mott in his address to the 1912 International Eugenics Congress, *Problems in Eugenics*, I, 405.
10. For example, A. F. Tredgold, 'The Study of Eugenics', *Quarterly Review*, 217 (1912), 50–1.
11. J. B. Haycraft, *Darwinism and Race Progress* (1895), pp. 62–3, 68.
12. Arnold Henry White, *Efficiency and Empire* [1901], edited with an introduction and notes by G. R. Searle, Society and the Victorians XV (Brighton, 1973), p. 110.
13. H. Giffard-Ruffe, 'A Plea for Posterity', *Westminster Review*, 156 (1901), p. 33.
14. *Weekly Sun*, 28 July 1900.
15. 'Miles', 'Where to Get Men', *Contemporary Review*, 81 (1902), 78–86; Bentley B. Gilbert, *The Evolution of National Insurance in Great Britain* (1966), pp. 84–6.
16. Cited in A. Watt Smyth, *Physical Deterioration: Its Causes and the Cure* (1904), pp. 13–14.
17. The arguments have been usefully summarized in ibid., pp. 14–17, 21–2, 297–301.
18. *Report of the Inter-Departmental Committee on Physical Deterioration*, Parliamentary Papers, 1904 [Cd. 2175], XXXII, 1.

19. For example, the *Report* was cited by the *Manchester Guardian*, 22 March 1910, in its public altercation with Pearson.
20. Havelock Ellis, *The Problem of Race-Regeneration* (1911), p. 30.
21. Jones, *Outcast London*, p. 127.
22. *Nature*, 72 (1905), 331.
23. Havelock Ellis, *The Task of Social Hygiene* (1912), p. 181.
24. G. A. Reid, 'Recent Researches in Alcoholism', *Bedrock*, 1 (1912), 37–8.
25. For example, Lt.-Col. W. Hill-Climo, cited in *ER*, 1 (1909–10), 72–3.
26. S. Webb, 'Physical Degeneracy or Race Suicide?', *The Times*, 16 October 1906.
27. A. E. Crawley, 'Primitive Eugenics', *ER*, 1 (1909–10), 275–80 (p. 278).
28. W. C. D. Whetham, *Eugenics and Unemployment* (Cambridge, 1910), p. 15.
29. National Birth-Rate Commission, *The Declining Birth-Rate* (1916), p. 9.
30. W. C. D. Whetham, 'Heredity and Destitution', *ER*, 3 (1911–12), 131–42 (p. 139).
31. National Birth-Rate Commission, *Problems of Population and Parenthood* (1920), p. xxxix.
32. David Heron, *On the Relation of Fertility in Man to Social Status, and on the Changes in this Relation that have taken Place during the Last Fifty Years*, Drapers' Company Research Memoirs, Studies in National Deterioration I (1906), p. 21.
33. Eugenists sometimes appealed to social reformers to co-operate with them by arguing that, *whatever view be taken of the importance of heredity*, the undesirability of a differential birth-rate could hardly be denied.
34. C. W. Saleeby, *The Progress of Eugenics* (1914), p. 27.
35. Heron, *Relation of Fertility to Social Status*, p. 21.
36. K. Pearson, *The Scope and Importance to the State of the Science of National Eugenics* (1909), p. 36.
37. A. F. Tredgold, 'The Feeble-Minded', *Contemporary Review*, 97 (1910), 721.
38. W. C. D. Whetham in a review of Archdall Reid's *Laws of Heredity*, *ER*, 3 (1911–12), 66–7.
39. W. C. D. Whetham & C. D. Whetham, *The Family and the Nation* (1909), pp. 148–57.
40. For example, K. Pearson, *Social Problems: Their Treatment, Past Present and Future* (1912), p. 38.
41. F. C. S. Schiller, 'Eugenics and Politics', *Hibbert Journal*, 12 (1913–14), 241–59 (p. 250).
42. Whethams, *Family and Nation*, p. 131.
43. Whethams, *Introduction to Eugenics*, p. 59; K. Pearson, *The Groundwork of Eugenics* (1909), p. 37.
44. George P. Mudge, 'Biological Iconoclasm, Mendelian Inheritance and Human Society', *Mendel Journal*, 1 (1909), 45–124 (p. 95).
45. Whethams, *Introduction to Eugenics*, p. 36.
46. *ER*, 2 (1910–11), 2.

47. W. C. D. Whetham & C. D. Whetham, 'Decadence and Civilisation', *Hibbert Journal*, 10 (1911–12), 179–200 (p. 186).
48. D. Heron, *A First Study of the Statistics of Insanity and the Inheritance of the Insane Diathesis*, Eugenics Laboratory Memoirs II (1907), p. 6.
49. Robert Reid Rentoul, *Race Culture: or Race Suicide?* (1906), p. xi.
50. K. Pearson, *Darwinism, Medical Progress and Eugenics* (1912), p. 23.
51. Havelock Ellis, *Social Hygiene*, p. 42, fn. 1.
52. Mark H. Haller, *Eugenics: Hereditarian Attitudes in American Thought* (New Brunswick, 1963), pp. 106–8. Arnold White made a modest contribution of his own to this morbid literary genre: *The Views of 'Vanoc': An Englishman's Outlook* (1910), pp. 275–6.
53. W. R. Inge, *Outspoken Essays: Second Series* (1927), p. 266; W. C. Sullivan, 'Eugenics and Crime', *ER*, 1 (1909–10), 116–7; E. J. Lidbetter, *Heredity and the Social Problem Group* (1933), p. 18.
54. A. F. Tredgold, 'The Study of Eugenics', *Quarterly Review*, 217 (1912), 59.
55. R. B. Cattell, *The Fight for our National Intelligence* (1937), p. 124.
56. Quoted in W. A. Chapple, *The Fertility of the Unfit* (Melbourne, [1904]), p. 91.
57. See Haller, *Eugenics*, pp. 15–16.
58. Charles Goring, *The English Convict: A Statistical Study* (1913), especially pp. 364, 370–4.
59. Havelock Ellis, *Race-Regeneration*, pp. 40–1.
60. Haller, *Eugenics*, chapter 7.
61. *Report of the Royal Commission on the Care and Control of the Feeble-Minded*, Parliamentary Papers, 1908 [Cd. 4202], XXXIX, 159, para. 589.
62. A. J. Balfour, 'Decadence' [1908], in *Essays Speculative and Political* (1920), pp. 1–52.
63. For example, A. J. Hubbard, *The Fate of Empires: being an Inquiry into the Stability of Civilisation* (1913). See the critical comments on this literature by Sidney Low: 'Is Our Civilisation Dying?', *Fortnightly Review*, 93 (1913), 627–39.
64. W. Bateson, 'Heredity in the Physiology of Nations', *The Speaker*, 14 October 1905.
65. Galton, *Essays*, pp. 38–9.
66. Pearson, *Groundwork of Eugenics*, p. 20.

Chapter 4

1. Quoted by C. W. Saleeby, *Parenthood and Race Culture* (1909), p. 201.
2. Edward Fry *et al.*, *The Problem of the Feeble-Minded* (1909), p. 85.
3. A. White, 'Eugenics and the National Efficiency', *ER*, 1 (1909–10), 105–11.
4. W. C. D. Whetham & C. D. Whetham, *The Family and the Nation* (1909), p. 9.
5. *ER*, 2 (1910–11), p. 150.
6. Galton's observations in the *Westminster Gazette*, 26 June 1908; K. Pearson, *The Academic Aspect of the Science of National Eugenics* (1911), p. 27; James

Barr, *The Aim and Scope of Eugenics* (Edinburgh, 1911), p. 11.
7. E. Alec-Tweedie, 'Eugenics', *Fortnightly Review*, 91 (1912), 856.
8. W. C. D. Whetham & C. D. Whetham, *An Introduction to Eugenics* (Cambridge, 1912), p. 36.
9. Galton, *Essays*, p. 34.
10. For example, S. Low, 'Is Our Civilisation Dying?', *Fortnightly Review*, 93 (1913), 630.
11. *Nature*, 91 (1913), 85.
12. National Birth-Rate Commission, *The Declining Birth-Rate* (1916), p. 74.
13. M. Crackanthorpe, *Population and Progress* (1907), especially p. 131; Galton to Pearson, 11 June 1907, in Pearson, *Life*, IIIA, 322.
14. K. Pearson, *National Life From the Standpoint of Science* (1901), pp. 43–4.
15. Col. C. H. Melville, 'Eugenics and Military Service', *ER*, 2 (1910–11), 54.
16. *Problems in Eugenics*, II, 49.
17. Barr, *Scope of Eugenics*, p. 17.
18. Mark H. Haller, *Eugenics: Hereditarian Attitudes in American Thought* (New Brunswick, 1963), p. 88. The author of the remark was Irving Fisher.
19. A. M. Carr-Saunders, *Eugenics* (1926), p. 190.
20. L. Darwin, *The Need for Eugenic Reform* (1926), pp. 499–504.
21. *Problems in Eugenics*, II, 47–9.
22. *ER*, 1 (1909–10), 72–3.
23. EES Minutes, 6 March 1912 and 11 July 1913.
24. 'Eugenics and the War', *ER*, 6 (1914–15), 195–203.
25. W. R. Inge, 'Depopulation', *ER*, 5 (1913–14), 261.
26. For example, J. W. Slaughter, 'Selection in Marriage', *ER*, I (1909–10), 150–62 (p. 155).
27. J. F. Tocher, 'The Necessity for a National Eugenic Survey', *ER*, 2 (1910–11), 124–41 (pp. 134–8). The attempt to explain particular types of achievement in terms of racial ancestry can also be found in F. Galton, *English Men of Science: Their Nature and Nurture* (1874), and in Havelock Ellis, *A Study of British Genius* (1904).
28. *Annals of Eugenics*, 1 (1925–6), 1.
29. For example, the Whethams' paper, 'The Influence of Race on History', delivered to the First International Eugenics Congress; see *Problems in Eugenics*, I, 237–46.
30. Haller, *Eugenics*, pp. 78–80; John Higham, *Strangers in the Land: Patterns of American Nativism, 1860–1925* (New York, 1963), passim.
31. A. F. Tredgold, 'Eugenics and the Future Progress of Man', *ER*, 3 (1911–12), 94–117 (p. 111).
32. Whethams' *Introduction to Eugenics*, pp. 43–4.
33. On the 'restrictionists', see Bernard Gainer, *The Alien Invasion: The Origins of the Aliens Act of 1905* (1972). For links with eugenists, see especially pp. 114–5.
34. K. Pearson & M. Moul, 'The Problem of Alien Immigration into Great Britain,

Illustrated by an Examination of Russian and Polish Jewish Children', *Annals of Eugenics,* 1 (1925–6), especially pp. 5–9, 126–7. The issue has been conveniently summarized in N. Pastore, *The Nature-Nurture Controversy* (New York, 1949), pp. 33–6.

35. EES Minutes, 14 March 1913.
36. K. Pearson, *The Scope and Importance to the State of the Science of National Eugenics* (1909), p. 44.
37. W. C. D. Whetham & C. D. Whetham, *Heredity and Society* (1912), pp. 46–7.
38. A. White, *The Views of 'Vanoc': An Englishman's Outlook* (1910), p. 284.
39. Redcliffe N. Salaman, 'Heredity and the Jew', *ER,* 3 (1911–12), 187–200 (pp. 197, 199). This paper was criticized by a prominent Jewish eugenist, Sidney Herbert, in the following number of the *Eugenics Review*: 3, no. 4 (January 1912), 349–51. Incidentally, the young Laski also wrote, at Pearson's instigation, an article refuting Salaman, which was later published in *Biometrika*; see [Basil] Kingsley Martin, *Harold Laski (1893–1950)* (1953), p. 11, fn. 4.
40. *ER,* 1 (1909–10), 222.
41. J. M. Winter, 'The Webbs and the Non-White World: A Case of Socialist Racialism', *Journal of Contemporary History,* 9, no. 1 (January 1974), 181–92.
42. W. McDougall, 'Psychology in the Service of Eugenics', *ER,* 5 (1913–14), 295–308 (pp. 306–7).
43. The tests applied to American soldiers during the First World War were frequently offered as evidence of Negro inferiority and of the superiority of the old 'Anglo-Saxon' stocks, although, of course, this interpretation was challenged even at the time.
44. EES Minutes, 1 February 1911. It is here reported that Professor Haddon had suggested books for the guidance of the South African gentleman.
45. For example, Whethams, *Family and Nation,* pp. 157–8.
46. See R. Ruggles Gates, *Heredity and Eugenics* (1923), passim.
47. For example, 'Vanoc', 'The Common Sense of Heredity: Devil or Saviour?', *Referee,* 6 September 1908.
48. 'Black and White Marriages. The New American Crusade: Scientific Views', *Observer,* 3 May 1908.
49. K. Pearson, *Social Problems: Their Treatment, Past Present and Future* (1912), p. 9.
50. A. E. Crawley, 'Primitive Eugenics', *ER,* 1 (1909–10), 275–80.
51. See W. Bateson, 'Presidential Address to the British Association Meeting, 1914', in *William Bateson: His Essays and Addresses, together with a Short Account of his Life by Beatrice Bateson* (1928), p. 310.
52. J. B. S. Haldane, *Heredity and Politics* (1938), p. 155.
53. For example, Reginald Ruggles Gates, a Canadian by birth who had spent some time at American universities before becoming Professor of Botany at King's College, London. His remarks about the evils of 'misceganation' rest heavily on American 'authorities'.

Notes and References

Chapter 5

1. S. Low, 'Darwinism and Politics', *Fortnightly Review*, 86 (1909), 527. Maurice Eder wrote ironically in the *New Age* for 16 December 1909: 'Professor Pearson will doubtless now call upon the State to house him, his staff, and the middle classes generally in the slums, and to provide palaces for the poor, whom he is so anxious to be rid of'.
2. S. Webb, 'Eugenics and the Poor Law: The Minority Report', *ER*, 2 (1910–11), 233–41 (p. 236).
3. W. C. D. Whetham & C. D. Whetham, *Heredity and Society* (1912), p. 8.
4. As Professor J. A. Thomson, a 'reform eugenist' argued: 'if individual modificational gains are not handed on, neither are the losses . . . Many detrimental acquired characters are to be seen all around us, but if they are not transmissible, they need not last'; see his *Darwinism and Human Life* (1909), p. 173.
5. K. Pearson, *Eugenics and Public Health* (1912), pp. 32–4.
6. L. Darwin, 'Our Critic Criticized', *ER*, 5 (1913–14), 316–42 (p. 322); Presidential Address, 1 June 1911, *EES Annual Report 1910–11*, pp. 9–10.
7. Interview with Pearson, *The Standard*, 13 November 1910.
8. K. Pearson, *The Problem of Practical Eugenics* (1909), p. 34.
9. Amy Barrington & K. Pearson, *A First Study of the Inheritance of Vision and of the Relative Influence of Heredity and Environment on Sight*, Eugenics Laboratory Memoirs V (1909), p. 61.
10. E. Schuster, 'The Scope of the Science of Eugenics', *British Medical Journal*, no vol. no. (2 August 1913), 223–5 (p. 225).
11. D. Heron, *The Influence of Defective Physique and Unfavourable Home Environment on the Intelligence of School Children . . .*, Eugenics Laboratory Memoirs VIII (1910), p. 4.
12. K. Pearson, *The Scope and Importance to the State of the Science of National Eugenics* (1909), p. 25.
13. L. Darwin, *Need for Eugenic Reform* (1926), p. 296.
14. L. Darwin, 'The Cost of Degeneracy: Being Part of the Annual Presidential Address', *ER*, 5 (1913–14), 93–100 (pp. 96–8).
15. ibid., pp. 94–5.
16. M. Crackanthorpe, 'Eugenics as a Social Force', *19th Century*, 63 (1908), 970.
17. *Sheffield Telegraph*, 31 March 1910.
18. W. Bateson, 'Presidential Address to the British Association Meeting, 1914', in *William Bateson: His Essays and Addresses, together with a Short Account of his Life by Beatrice Bateson* (1928), p. 309.
19. ibid., p. 338.
20. Whethams, *Heredity and Society*, p. 126.
21. Editorial, *Annals of Eugenics*, 1 (1925–6), 3.
22. E. J. Lidbetter, 'Nature and Nurture—A Study in Conditions', *ER*, 4 (1912–13), 54–73 (pp. 72–3).

23. F. Galton, *Hereditary Genius* (1869), pp. 33–43.
24. F. Galton, *English Men of Science: Their Nature and Nurture* (1874), p. 23.
25. W. McDougall, 'A Practical Eugenic Suggestion', *Sociological Papers*, 3 (1907), 63–4.
26. Cyril Burt, 'Experimental Tests of General Intelligence', *British Journal of Psychology*, 3 (1909), 94–177 (especially pp. 175–6).
27. W. McDougall, 'Psychology in the Service of Eugenics', *ER*, 5 (1913–14), 295–308; C. Burt, 'The Measurement of Intelligence by the Binet Tests', *ER*, 6 (1914–15), 36–50, 140–52; C. Spearman, 'The Measurement of Intelligence', *ER*, 6 (1914–15), 312–13; T. Simon, 'The Measurement of Intelligence', *ER*, 6 (1914–15), 291–307.
28. Darwin, *Eugenic Reform*, p. 271.
29. Presidential Address, 1 June 1911, *EES Annual Report, 1910–11*, p. 14.
30. Bateson, *Essays*, pp. 354–5.
31. G. P. Mudge, 'Plea for a More Virile Sentiment in Human Affairs', *Mendel Journal*, 1 (1909), 60–1. See also M. W. Keatinge, 'Education and Eugenics', *ER*, 6 (1914–15), 97–115 (p. 105).
32. For example, F. C. S. Schiller's remarks at the First International Eugenic Congress, *Problems in Eugenics*, II, 43–4.
33. Darwin, *Eugenic Reform*, p. 157.
34. W. C. D. Whetham & C. D. Whetham, *The Family and the Nation* (1909), pp. 191–2.
35. Darwin, *Eugenic Reform*, p. 455–6.
36. J. F. Tocher, 'The Necessity for a National Eugenic Survey', *ER*, 2 (1910–11), 124.
37. Whethams, *Heredity and Society*, p. 142.
38. Whethams, *Family and Nation*, p. 222.
39. Presidential Address, 5 May 1910, *ESS Annual Report, 1909–10*, p. 6.
40. K. Pearson, *On the Handicapping of the First Born* (1914), pp. 67–8.
41. ibid., passim. In the 1920s, when birth control was much more widely acceptable, some eugenists claimed that there was evidence that the 'youngest children are generally imperfectly developed, feeble in growth, and often deformed in structure', thereby standing Pearson's arguments on their head. See E. W. MacBride, 'Social Biology and Birth-Control', *Nature* 113 (1924), 773–5.
42. *The Times*, 21 March 1910.
43. *Nature*, 86 (1911), 485.
44. In his testimony before the Royal Commission on Income Tax in 1919, Major Darwin suggested that it might be desirable to differentiate between inherited wealth, and earned income and interest on the investment of savings. Wealth had a eugenic meaning, in so far as it represented money which had been 'earned, saved, and not squandered uselessly'. But *inherited* wealth was another matter, and could, he suggested, be taxed at a higher rate, since it was not a reward for efficiency. See *Royal Commission on Income Tax: Minutes of*

Evidence, Appendices and Index, Parliamentary Papers, 1919 [Cmd. 288], XXIII, pt I, p. 785.

45. J. B. Haycraft, *Darwinism and Race Progress* (1895), p. 123.
46. F. C. S. Schiller, 'Practicable Eugenics in Education', *Problems in Eugenics*, I, 164–5, 170.
47. Whethams, *Family and Nation*, p. 118.
48. See Shaw's address at the City Temple Literary and Debating Society on 'Christian Economics', 30 October 1913, in A. Chappelow, *Shaw—'The Chucker Out'* (1969), especially p. 148.
49. A. R. Wallace, *Social Environment and Moral Progress* (1913), passim.
50. C. W. Saleeby, *Parenthood and Race Culture* (1909), pp. 194–5.
51. Darwin, *Eugenic Reform*, p. 363.
52. Bateson, *Essays*, p. 315.
53. F. C. S. Schiller, 'Practicable Eugenics in Education', *Problems in Eugenics*, I, 162–71. For a similar expression of opinion, voiced at a later date, see R. A. Fisher, 'Family Allowances in the Contemporary Economic Situation', *ER*, 24 (1932–3), 87–95 (p. 90).
54. Galton, *Hereditary Genius*, p. 348.
55. For example, the Rev. Edward Lyttelton, Headmaster of Eton, discussing the falling birth-rate, proudly announced that he had gathered some interesting and relevant material from 'a man in close contact with the working classes, who was able to make friends with a large number of the higher class of artisan in a country district'; see *ER*, 5 (1913–14), 35.
56. R. J. Halliday, 'Social Darwinism: A Definition', *Victorian Studies*, 14 (1971) 404.
57. E. J. Lidbetter, 'Nature and Nurture—A Study in Conditions', *ER*, 4 (1912–13), 69.
58. Darwin, *Eugenic Reform*, p. 307.
59. See Darwin's distinction between the 'naturally unfit' and 'the unlucky' in his Presidential Address, 20 June 1912, *EES Annual Report, 1911–12*, p. 6.
60. W. C. D. Whetham, *Eugenics and Unemployment* (Cambridge, 1910), p. 12.
61. Galton to Pearson, 12 July 1907; Pearson to Galton, July (?) 1907; quoted in Pearson, *Life*, IIIA, 323.
62. EES Minutes, 3 November 1909 and 12 January 1910.
63. E. J. Lidbetter, 'Some Examples of Poor Law Eugenics', *ER*, 2 (1910–11), 204–28.
64. 'Report of the Committee on Poor Law Reform: Section I . . .', *ER*, 2 (1910–11), 167–77; E. J. Lidbetter, 'Eugenics and the Prevention of Destitution', *ER*, 3 (1911–12), 170–3.
65. Lidbetter, 'Some Examples of Poor Law Eugenics', p. 206.
66. Edward Brabrook, 'Eugenics and Pauperism', *ER*, 1 (1909–10), 229–41 (pp. 231–2).
67. Lidbetter, 'Some Examples of Poor Law Eugenics', p. 223.
68. Mark H. Haller, *Eugenics: Hereditarian Attitudes in American Thought* (New

Brunswick, 1963), p. 104.
69. K. Pearson, *Mendelism and the Problem of Mental Defect: III. . .*, Questions of the Day and of the Fray No. 9 (1914), p. 50.
70. 'Report of the Committee on Poor Law Reform: Section I . . .', p. 173.
71. Brabrook, 'Eugenics and Pauperism', p. 232.
72. 'Report of the Committee on Poor Law Reform: Section III . . .', *ER*, 2 (1910–11), 186–94 (p. 193).
73. Lidbetter, 'Some Examples of Poor Law Eugenics', p. 223.
74. Sidney & Beatrice Webb, *The Prevention of Destitution* (1911), p. 46.
75. C. S. Loch, 'Eugenics and the Poor Law: The Majority Report', *ER*, 2 (1910–11); 229–32 (p. 230).
76. See Bernard Bosanquet, 'The Problem of Selection in Human Society', *Charity Organisation Review*, 28 (1910), 369–86.
77. Whethams, *Family and Nation*, pp. 171–2.
78. Pearson, *Problem of Practical Eugenics*, pp. 31, 35.
79. EES Minutes, 4 October 1912.
80. C. W. Saleeby, *The Progress of Eugenics* (1914), p. 30.
81. See J. H. Koeppern, 'Maternity Insurance', *ER*, 1 (1909–10), 281–3.
82. A. C. Gotto, 'The Relation of Eugenic Education to Public Health', *Journal of State Medicine*, 21 (1013), 625.
83. EES Minutes, 30 May 1913.

Chapter 6

1. C. W. Saleeby, *Parenthood and Race Culture* (1909), p. viii.
2. *William Bateson: His Essays and Addresses, together with a Short Account of his Life by Beatrice Bateson* (1928), pp. 14–16.
3. J. B. Haycraft, *Darwinism and Race Progress* (1895), pp. 13–14.
4. A. F. Tredgold, 'Heredity and Environment in Regard to Social Reform', *Quarterly Review*, 219 (1913), 382–3.
5. W. R. Inge, *The Church and the Age* (1912), p. 15.
6. A. White, *The Problems of a Great City* (1886), p. 27.
7. Pearson, *Life*, IIIA, 349.
8. Ethel M. Elderton, *et al.*, *On the Correlation of Fertility with Social Value . . .*, Eugenics Laboratory Memoirs No. XVII (1913), p. 46.
9. W. C. D. Whetham, 'Inheritance and Sociology', *19th Century*, 65 (1909), 87.
10. J. W. Slaughter, 'Selection in Marriage', *ER*, 1 (1909–10), 161.
11. J. A. Lindsay, 'The Case For and Against Eugenics', *19th Century*, 72 (1912), 554.
12. Bateson, *Essays*, p. 344, fn. 1.
13. *Manchester Guardian*, 22 March 1910.
14. K. Pearson, *The Scope and Importance to the State of the Science of National Eugenics* (1909), passim.
15. K. Pearson, *Nature and Nurture: the Problem of the Future* (1910), pp. 28–9.

16. K. Pearson, *The Academic Aspect of the Science of National Eugenics* (1911), p. 20.
17. For example, James Barr, *The Aim and Scope of Eugenics* (Edinburgh, 1911), p. 18. On other occasions Barr singled out Lloyd George in person for abuse; see Sir James Barr, 'Address to the Section of Child Study and Eugenics at the Dublin Congress', *Journal of Royal Institute of Public Health*, 19 (1911), 715.
18. Saleeby, *Parenthood*, pp. 118–9.
19. C. W. Saleeby, *Woman and Womanhood* (1912), cited *Sociological Review*, 6 (1913), 367; C. W. Saleeby, 'Racial Poisons: II. Alcohol', *ER*, 2 (1910–11), 34.
20. F. C. S. Schiller, *Eugenics and Politics* (1926), pp. 195–6.
21. F. C. S. Schiller, 'Eugenics and Politics', *Hibbert Journal*, 12 (1913–14), 252.
22. *Daily Sketch*, 9 February 1910.
23. EES Minutes, 12 January 1910.
24. ibid., 2 March 1910.
25. ibid., 1 November 1911.
26. ibid., 17 July 1914.
27. ibid., 19 December 1913.

Chapter 7

1. Galton, *Essays*, p. 24.
2. ibid., p. 100.
3. R. A. Fisher, paper on 'Heredity', delivered 10 November 1911; see Cambridge University Eugenics Society Records.
4. Beatrice Webb's *Our Partnership*, edited by M. Cole (1948), p. 257, journal entry 16 January 1903.
5. Quoted in Pearson, *Life*, II, 365.
6. *Problems in Eugenics*, I, 5.
7. K. Pearson, *The Groundwork of Eugenics* (1909), p. 19.
8. Galton, *Essays*, pp. 37, 104.
9. ibid., p. 63.
10. *EES Annual Report, 1908*, p. 10.
11. Report J. Ewart, 'The Influence of Parental Age on Offspring', *ER*, 3 (1911–12), 201–32 (p. 204).
12. Galton, *Essays*, p. 66.
13. Ewart, 'The Influence of Parental Age . . .', pp. 222–3.
14. James Barr, 'Address to the Section of Child Study and Eugenics at the Dublin Congress', *Institute of Public Health*, 19 (1911), 708.
15. As eugenists often pointed out; see C. W. Saleeby, 'Selecting Our Parents: The New Imperialism That Preaches the Necessity of Race Culture', *Sunday Chronicle*, 8 March 1908.
16. K. Pearson, *On the Relationship of Health to the Psychical and Physical Characters in School Children* (1923), pp. 59–60.
17. For example, K. Pearson, *The Problem of Practical Eugenics* (1909), p. 33.

18. Galton, *Essays*, p. 104.
19. *EES Annual Report, 1911–12*, p. 13.
20. *William Bateson: His Essays and Addresses, together with a Short Account of his Life by Beatrice Bateson* (1928), pp. 375, 377.
21. J. A. Lindsay, 'The Case For and Against Eugenics', *19th Century*, 72 (1912), 549, 553.
22. See James Joll, 'The English, Friedrich Nietzsche and the First World War', in *Deutschland in der Weltpolitik des 19 und 20 Jahrhunderts*, edited by I. Geiss & B. J. Wendt (Düsseldorf, 1973), pp. 287–305. However, the author misses the significance of the eugenic movement.
23. M. D. Eder, 'Good Breeding or Eugenics', *New Age*, 23 May 1908. See also the many attempts of M. A. Mügge to portray Nietzsche as a pioneer of eugenics; e.g., *Friedrich Nietzsche* (n.d. [1914?]), pp. 77–83.
24. G. Lowes Dickinson, *Justice and Liberty: A Political Dialogue* (1908), p. 46.
25. Benjamin Kidd, *Principles of Western Civilisation*, second edition (1908), p. xviii.
26. *Daily News*, 27 May 1910.
27. Galton, *Essays*, p. 35.
28. Franklin Kidd, letter to *The Times*, 16 October 1913.
29. L. T. Hobhouse, *Social Evolution and Political Theory* (New York, 1911), pp. 12, 42–3.
30. Urwick's review of *Parenthood and Race Culture* in *Sociological Review*, 3 (1910), 70.
31. F. C. S. Schiller, 'National Self-Selection', *ER*, 2 (1910–11), 8–24 (pp. 17, 24). See also F. C. S. Schiller, *Eugenics and Politics* (1926), pp. 94–7.
32. 'The Dangers of Eugenics', *The Nation*, 13 March 1909, p. 887.
33. Hobhouse, *Social Evolution*, pp. 69–70.
34. H. G. Wells, *Mankind in the Making* (1903), pp. 40–50. The point had earlier been made by T. H. Huxley in his attack on what he called 'this pigeon-fanciers' polity'; see his *Evolution and Ethics and Other Essays* (1894), pp. 22–3.
35. C. W. Saleeby, *The Progress of Eugenics* (1914), p. 174.
36. C. W. Saleeby, 'The Obstacles to Eugenics', *Sociological Review*, 2 (1909), 229.
37. Galton, *Essays*, pp. 100–9.
38. J. F. Tocher, 'The Necessity for a National Eugenic Survey', *ER*, 2 (1910–11), 136–7.
39. R. A. Fisher, 'Some Hopes of a Eugenist', delivered 22 November 1912, Cambridge University Eugenics Society Records.
40. Galton, *Essays*, pp. 24–5, 30–2.
41. C. P. Blacker, *Eugenics: Galton and After* (1952), p. 122.
42. Galton, *Essays*, pp. 21–2.
43. ibid., pp. 31–2.
44. F. Galton, 'Hereditary Improvement', *Frazer's Magazine*, 7 (1873), 125.

45. W. C. D. Whetham, *Eugenics and Unemployment* (Cambridge, 1910), p. 20.
46. Galton, 'Hereditary Improvement', p. 123.
47. ibid., pp. 127–9.
48. C. W. Armstrong, *The Survival of the Unfittest* (1927), chapter 9. In the 1930s, he was writing about possible eugenic settlements in Brazil; see his 'A Eugenic Colony: A Proposal for South America', *ER*, 25 (1933–4), 91–7.
49. Anon., *Essays in Buff* (1903). See also 'The Author of "Essays in Buff"', 'Thoughts on Eugenics', *ER*, 3 (1911–12), 233–8.
50. L. Darwin, *The Need for Eugenic Reform* (1926), especially pp. 157, 166.
51. Galton, *Essays*, pp. 63–4.
52. Havelock Ellis, *The Task of Social Hygiene* (1912), pp. 202–3.
53. J. W. Slaughter, 'Selection in Marriage', *ER*, 1 (1909–10), 150–62 (p. 153).
54. Havelock Ellis, *Social Hygiene*, p. 203, fn. 1.
55. *Daily Sketch*, 22 March 1910.
56. W. C. D. Whetham & C. D. Whetham, *The Family and the Nation* (1909), p. 201.
57. *EES Annual Report, 1909–10*, p. 15.
58. J. A. Lindsay, 'The Case For and Against Eugenics', *19th Century*, 72 (1912), 552.
59. Lowes Dickinson, *Justice and Liberty*, pp. 137–8.
60. Sidney Herbert, 'Eugenics in Relation to Social Reform', *Westminster Review*, 180 (1913), 384.
61. For example, C. J. Bond suggested at a later date that financial help could be restricted to the first four or five children, in order 'to reduce the number of children in the too large families in the lower groups', while at the same time increasing 'the, at present, too small families in the middle section of Society'; see 'A Surgeon' [C. J. Bond], *Essays and Addresses: Sociological, Biological and Psychological* (1930), pp. 143–4.
62. W. C. D. Whetham & C. D. Whetham, *Heredity and Society* (1912), pp. 95–7.
63. *ER*, 1 (1909–10), 225–6.
64. Whethams, *Heredity and Society*, pp. 159–60.
65. C. W. Saleeby, *Parenthood and Race Culture* (1909), p. 151.
66. Saleeby, *Progress of Eugenics*, pp. 67–8.
67. Whethams, *Family and Nation*, p. 91.
68. Pearson, *Problem of Practical Eugenics*, p. 30.
69. W. McDougall, 'A Practical Eugenic Suggestion', *Sociological Papers*, 2 (1907), 76–9.
70. Pearson, *Problem of Practical Eugenics*, p. 30.
71. Beatrice & Sidney Webb, 'What is Socialism?: XIII. Freedom for the Woman', *New Statesman*, 5 July 1913.
62. Pearson, *Problem of Practical Eugenics*, p. 29.
73. EES Minutes, 1 June 1911.
74. National Maritime Museum, Greenwich, Arnold White Papers, file 55, memo of 2 June 1914.

75. *Royal Commission on Income Tax: Minutes of Evidence, Appendices and Index*, Parliamentary Papers, 1919 [Cmd. 288], XXIII. pt I, pp. 783–88.
76. *Hansard*, 5th series (Commons), 65, cols 523–6, 22 July 1914.
77. *Royal Commission on Income Tax: Report*, Parliamentary Papers, 1920 [Cmd. 615], XVIII, 97, para. 270.
78. *Royal Commission on Income Tax: Minutes . . .*, pp. 783–5.
79. *Problems in Eugenics*, II, 29.
80. *Royal Commission on Income Tax: Minutes . . .*, p. 784 and qu. 15, 854.
81. *Royal Commission on Income Tax: Report*, paras, 263–4.

Chapter 8

1. W. A. Chapple, *The Fertility of the Unfit* (Melbourne, [1904]), p. 102. Th book cited is Dr McKim's *Heredity and Human Progress* (1900).
2. *Dail News*, 9 November 1909.
3. *Daily Express*, 4 March 1910.
4. For example, R. R. Rentoul, 'Sterilisation of Certain Degenerates', *Public Health Review*, February 1910; and his 'Sterilising the Insane', *ER*, 2 (1910–11), 74–6.
5. Macleod Yearsley, 'Eugenics and Congenital Deaf-Mutism', *ER*, 2 (1910–11), 299–312 (pp. 311–12).
6. A. White, *The Great Idea*, Reports on the Social Work of the Salvation Army 1909/10 (1909–10), p. 17.
7. A. White, *The Views of 'Vanoc': An Englishman's Outlook* (1910), pp. 285, 281.
8. J. A. Lindsay, 'The Case For and Against Eugenics', *19th Century*, 72 (1912), 557.
9. Paper delivered 27 October 1911, Cambridge University Eugenics Society Records.
10. Mark H. Haller, *Eugenics: Hereditarian Attitudes in American Thought* (New Brunswick, 1963), pp. 49–50, 209, fn. 33.
11. ibid., pp. 130–41.
12. J. B. S. Haldane, *Heredity and Politics* (1938), p. 80.
13. Havelock Ellis, 'The Sterilisation of the Unfit', *ER*, 1 (1909–10), 203–6; and *The Problem of Race-Regeneration* (1911), pp. 66–7.
14. Anon, *Essays in Buff* (1903), p. 110.
15. C. T. Ewart, 'Eugenics and Degeneracy', *Journal of Mental Science*, 56 (1910), 682; E. Faulks, 'The Sterilisation of the Insane', ibid., 57 (1911), 63–74; Geoffrey Clarke, 'Sterilisation from the Eugenic Standpoint . . .', ibid., 58 (1912), 48–61.
16. C. W. Saleeby, *The Progress of Eugenics* (1914), pp. 208–9.
17. L. Darwin, *The Need for Eugenic Reform* (1926), p. 176.
18. For example, F. W. Mott's observations; see *Problems in Eugenics*, I, 428.
19. Havelock Ellis, *Race-Regeneration*, pp. 68–9.

20. A. F. Tredgold, 'The Study of Eugenics', *Quarterly Review*, 217 (1912), 63–4.
21. E. Alec-Tweedie, 'Eugenics', *Fortnightly Review*, 91 (1912), 859.
22. F. C. S. Schiller, *Eugenics and Politics* (1926), p. 152.
23. 'Report of the Committee on Poor Law Reform: Section I . . .', *ER*, 2 (1910–11), 177. My italics.
24. A. White, *The Problems of a Great City* (1886), p. 48. See also A. White, *Efficiency and Empire* (1901), p. 120.
25. J. W. Slaughter, 'Selection in Marriage', *ER*, 1 (1909–10), 158.
26. M. Crackanthorpe, *Population and Progress* (1907), p. 93.
27. W. C. D. Whetham, *Eugenics and Unemployment* (Cambridge, 1910), p. 18.
28. S. Squire Sprigge, 'Mating and Medicine', *Contemporary Review*, 96 (1909), 578–87.
29. *ER*, 1 (1909–10), 214.
30. R. R. Rentoul, *Race Culture or Race Suicide?* (1906), p. 122.
31. *ER*, 6 (1914–15), 156.
32. See Shaw's address at City Temple, 30 October 1913, in A. Chappelow, *Shaw—'The Chucker Out'* (1969), pp. 145–8.
33. Galton, *Essays*, pp. 28–9.
34. ibid., p. 67.
35. ibid., pp. 44–59.
36. Saleeby, *Progress of Eugenics*, p. 106.
37. Schiller, *Eugenics and Politics*, p. 216.
38. Chappelow, *Shaw*, pp. 145–8.
39. Quoted by Havelock Ellis, 'The Cry of the Unborn', *New Age*, 11 April 1908.
40. Harry Campbell, 'Eugenics From the Physician's Standpoint', *British Medical Journal*, no vol. no. (2 August 1913), 225–7 (p. 226).
41. Anon., *Heredity and Disease* (1908).
42. EES Minutes, 14 February 1912.
43. Squire Sprigge, 'Mating and Medicine', p. 585.
44. Haldane, *Heredity and Politics*, p. 99.
45. K. Pearson, *Eugenics and Public Health* (1912), p. 4. Arthur Newsholme expressed the hostility of many Medical Officers of Health to eugenics, which he regarded, with some reason, as a fundamental attack on his life's work; see his *The Last Thirty Years in Public Health* (1936), pp. 208–11.
46. For example, L. Darwin's Presidential Address of 20 June 1912, *EES Annual Report, 1911–12*, p. 18.
47. *Report of the Departmental Committee on Sterilisation*, Parliamentary Papers, 1933–4 [Cmd. 4485], XV, 611, pp. 6–7.
48. K. Pearson, *The Problem of Practical Eugenics* (1909), p. 17. Havelock Ellis also attached great importance to 'Neo-Malthusian methods'; see his *The Task of Social Hygiene* (1912), pp. 163–4.
49. National Birth-Rate Commission, *The Declining Birth-Rate* (1916), p. 21.
50. See discussion in H. J. Habakkuk, *Population Growth and Economic Development* (Leicester, 1971), pp. 56, 67.

51. W. R. Inge, 'Depopulation', *ER*, 5 (1913–14), 262.
52. Crackanthorpe, *Population and Progress*, especially chapter 1.
53. White, *Problems of a Great City*, p. 58.
54. A. Newsholme, *The Declining Birth-Rate: Its National and International Significance* (1911), p. 64.
55. Ethel M. Elderton, *Report on the English Birthrate: Part I, England, North of the Humber* (1914), especially pp. 238–9.
56. EES Minutes, 3 February 1909 and 26 July 1912.
57. ibid., 6 July 1910.
58. James Barr, *The Aim and Scope of Eugenics* (Edinburgh, 1911), p. 4.
59. See the Memorandum that Drysdale submitted to the National Birth-Rate Commission in 1913, and printed in the Commission's *The Declining Birth-Rate* (1916), pp. 87–101.
60. J. Peel, 'Contraception and the Medical Profession', *Population Studies*, 18 (1964), 136–7.
61. ibid., p. 136.
62. In the spring of 1959, the Eugenics Society received various legacies under the will of Dr Marie Stopes, who had been a Life Fellow of the Society and a personal friend of its Secretary, Dr C. P. Blacker; see Faith Schenk & A. S. Parkes, 'The Activities of the Eugenics Society', *ER*, 60 (1968), 153. On developments in America, see Haller, *Eugenics*, pp. 88–92.
63. R. B. Cattell, *The Fight for our National Intelligence* (1937), p. 129.
64. *Problems in Eugenics*, II, 33. This observation was made at the First International Eugenics Congress in 1912, and could, perhaps, be discounted as a strategic ploy for gaining converts to his cause. But, significantly, he also laid considerable stress on eugenics in his evidence to the Birth-Rate Commission. Here he argued that, given another seven or eight years of progress, the quantity question would have been overcome and Neo-Malthusianism could then become the servant of 'negative eugenics'. His other main point was that, although family restriction may have been anti-eugenic in the past, this was because the better educated had taken advantage of Neo-Malthusian knowledge for themselves, but had put obstacles in the way of such knowledge reaching the poor. Drysdale reckoned that this state of affairs would soon be changed, to the benefit of the race; see National Birth-Rate Commission, *The Declining Birth-Rate*, especially pp. 89, 91.
65. Galton, *Essays*, p. 62.
66. Chapple, *Fertility of the Unfit*, 70–3.
67. A. M. Carr-Saunders, *Eugenics* (1926), p. 184.
68. White, *Problems of a Great City*, p. 50.
69. See L. Darwin, 'The Habitual Criminal', *ER*, 6 (1914–15), 204–18 (p. 210).

Chapter 9

1. This was claimed by McKenna himself; see *Hansard*, 5th series (Commons),

39, col. 628, 10 June 1912.

2. *The Times*, 11 August 1908.
3. *Report of the Royal Commission on the Care and Control of the Feeble-Minded*, Parliamentary Papers, 1908 [Cd. 4202], XXXIX, 159.
4. *Hansard*, 5th series (Commons), 1, col. 424, 22 February 1909.
5. ibid., 17, col. 1166, 13 June 1910.
6. *EES Annual Report, 1909–10*, p. 9.
7. Bodleian Library, Asquith Papers, Box 12, Churchill to Asquith, December 1910.
8. Cab/37/108/189.
9. *ER*, 2 (1910–11), 163–4.
10. W. S. Blunt, *My Diaries: Being a Personal Narrative of Events, 1888–1914*, 2 vols (1919–20), II, 416, 20 October 1912.
11. EES Minutes, 5 October 1910.
12. ibid., 1 March 1911.
13. ibid., 3 May 1911 and 1 June 1911.
14. EES Minutes, 28 November 1911; *ER*, 3 (1911–12), 355–8.
15. EES Minutes, 6 March 1912.
16. *Hansard*, 5th series (Commons), 38, cols. 1443–50, 17 May 1912.
17. ibid., cols 1461–2.
18. ibid., 45, col. 782, 12 December 1912.
19. For example, 'A National Peril: The Menace of the Feeble-Minded', reported in *Wolverhampton Express and Star*, 9 April 1910; and Mrs Pinsent's address in the Guildhall, Cambridge, *Cambridge Daily News*, 29 April 1912.
20. *The Times*, 28 November 1912.
21. *Hansard*, 5th series (Commons), 41, col. 710, 19 July 1912.
22. ibid., 38, col. 1474, 17 May 1912. The remark being quoted was actually made during the discussion of Mr Stewart's Private Member's Bill, but it was a theme to which Wedgwood repeatedly returned.
23. G. K. Chesterton's views are wittily and forcefully argued in *Eugenics and Other Evils* (1921).
24. See Chapple's unsuccessful move on an amendment to make marriage with a defective null and void.
25. '. . . we have also omitted any reference to what might be regarded as the Eugenic idea, which my hon. Friend behind me believes underlies the whole promotion of this Bill. I can assure him that, as the measure now stands, it exists for the protection of individual sufferers . . .'; *Hansard*, 5th series (Commons), 53, col. 221, 28 May 1913.
26. See Wedgwood's altercation with Chapple; ibid., 56, col. 84, 28 July 1913.
27. The EES Minutes, 19 March 1913, record that 'the New Mental Deficiency Act was in process of being drafted, Dr Langdon Down being absent from the Council in order to confer with the Home Office on some of the clauses'. Dr Langdon Down, who played a leading part in this campaign, was also a member of the National Association.

28. *ER*, 6 (1914–15), 52.
29. ibid., pp. 52–3, 55. The former Bill made it *compulsory* for local authorities to set up special schools for defective children. The second, extended restraint over inebriates, whether or not they were disorderly or criminal.

Chapter 10

1. John Rex, 'Nature versus Nurture: The Significance of the Revived Debate', in *Race, Culture and Intelligence*, edited by K. Richardson & D. Spears (1972), p. 178.
2. W. R. Inge, *Outspoken Essays: First Series* (1919), p. 101.
3. In the 1920s, the Society sent out a special appeal to '10,000 businessmen' (*Annual Report, 1926–7*)—to very little avail.
4. For a very clear statement of this position, see Eden Paul, *Socialism and Eugenics* (Manchester, 1911).
5. In the campaign for a voluntary sterilization Bill in the 1930s, the eugenists managed to win some significant support from members of the Labour Party, but were engaged in all-out war with the Catholic Church. The issuing of the Papal encyclical, *Casti Conubii*, in 1931 released the eugenists from any effort to show politeness towards the Catholic Church.
6. The possibilities of relating social and personal experience to scientific positions have been explored in an interesting monograph by Nicholas Pastore, *The Nature-Nurture Controversy* (New York, 1949).
7. For example, Mark H. Haller, *Eugenics: Hereditarian Attitudes in American Thought* (New Brunswick, 1963), p. 77.
8. ibid., p. 178.

Index

Index

Index

Index

Index